PRAISE FOR *ECO-ANXIETY*

"In this book, Heather White shows you how to reduce your eco-anxiety through service. Find your climate 'why' and learn how to apply it to environmental action in this easy-to-read, entertaining, and informative book on creating a greener, healthier future."

—DR. MARK HYMAN, *New York Times* bestselling author and head of strategy and innovation at the Cleveland Clinic Center for Functional Medicine

"Eco-anxiety is all too real for Generation Z. It's time for other generations to step up, partner with us, and help save our shared future. In *Eco-Anxiety: Saving Our Sanity, Our Kids, and Our Future*, Heather White shows you how to get involved in the movement by focusing on your strengths and celebrating a positive vision of what we can create together."

—MAYA PENN, Gen Z activist, author, artist, 3x TED speaker, and entrepreneur

"*Eco-Anxiety* doesn't dwell in gloom and doom but instead gives us just what the climate movement needs: a down-to-earth and back-to-joy approach to climate action."

—CHRIS HILL, Sierra Club's Chief Conservation Officer

"*Eco-Anxiety: Saving Our Sanity, Our Kids, and Our Future* is a joyful must-read for everyone—from the curious to the committed in the climate movement."

—COLLEEN WACHOB, cofounder and co-CEO of mindbodygreen

"In my faith tradition, a good person leaves an inheritance to their children's children. It's time to take a hard look at what kind of inheritance we are leaving the next generation. The world is facing a climate crisis an[illegible] help to avoid a

global calamity. In this book, Heather White creates a welcoming path for all to engage in daily practices of sustainability, to seek policy solutions, and to let our kids and grandkids know they will not face this crisis alone."

—LARRY SCHWEIGER, award-winning author, speaker, and former president and CEO of the National Wildlife Federation

"In this essential climate action handbook, Heather leverages her twenty years of experience in environmental advocacy and her warm, smart, and relatable approach to welcome you into the movement. She shows you how to make meaningful, lasting change and create a positive legacy for the next generation."

—GREGG RENFREW, founder & CEO of Beautycounter

"Heather continues to be a beacon of light and compass in helping families navigate climate change. Her work and network are precious resources that we can leverage to create change for future generations."

—ROBYN O'BRIEN, food activist and author

"Heather has created a positive, innovative, and engaging book to guide you on your climate action journey. It's a "choose-your-own adventure" environmental action book where you'll discover how to use your talents to create a better future for us all."

—DAN WENK, retired superintendent of Yellowstone National Park

ECO-ANXIETY

SAVING OUR SANITY, OUR KIDS, AND OUR FUTURE

HEATHER WHITE

Eco-Anxiety: Saving Our Sanity, Our Kids, and Our Future

Published by Harper Horizon, an imprint of HarperCollins Focus LLC.

All Eco-Hero interviews were conducted by the author in 2021 and are used with permission.

Graphic design by Molly Stratton (www.mollystratton.com)

ISBN 978-0-7852-9130-5 (eBook)
ISBN 978-0-7852-9129-9 (HC)
ISBN 978-0-7852-9132-9 (SC)

Library of Congress Control Number: 2021949068

Printed in the United States of America

24 25 26 27 28 LBC 5 4 3 2 1

To David, Cady, & Susan—
my favorite Wonk, Beacon, & Spark

CONTENTS

PART 2: THE SEVEN AREAS TO EFFECT CHANGE

PART 3: HOW IT ENDS

FOREWORD

As I write this foreword, oppressive heat waves, epic flooding, horrible droughts, catastrophic wildfires, melting glaciers, and fatal mudslides are affecting communities around the world. The list of extreme weather events goes on and on. The earth is begging for our help, and our kids and grandkids are understandably anxious and worried about the future they're inheriting. It's intense. Even for someone like me, who's dedicated my whole life to environmental action, the climate crisis can be so overwhelming it's hard to figure out where to start.

The pressing question is, Now what?

Read this book.

In *Eco-Anxiety: Saving Our Sanity, Our Kids, and Our Future* my friend Heather White will show you how to begin, first by identifying your key strengths in service—what she calls your Service Superpower—and then by helping you take action to support policy and market solutions. Heather walks you through how your individual actions impact the culture around you. That's when big things happen. Groups of people changing how they do things; pushing for more sustainable practices, clean energy, and safer products from companies; and demanding that their elected leaders act to protect our health and our environment. This creates lasting impact.

Whether she's engaged in lobbying on Capitol Hill for better environmental policy, doing a TV interview, or organizing concerned citizens, Heather knows what she's talking about. We've known each other for more than a decade, and I'm proud to have had her in my corner. Heather and I advocated for medical monitoring of veterans who were exposed to contaminated drinking water in Camp LeJeune, North Carolina. We fought to get federal agencies that register cancer diagnoses to work together to track how childhood exposures to toxic chemicals can make people sick later in life. We've also urged the federal government to clean up municipal drinking water systems and regulate contaminants. Heather is a smart, tenacious lawyer and policy expert, a warrior mom, and a champion for people and the planet.

If you're familiar with my work, you know that I love fighting for the underdog, standing up for people, and protecting the environment. As you begin your journey with this book, you'll find that it's not only empowering but also fun. Sometimes you learn shocking things when you take a deep dive into environmental health and climate policy, but the biggest wonder of it all is our humanity's resilience and knowing that we can make a difference, together.

In this book, Heather provides insights into your unique gifts, the ways you can show up for this movement, and the actions you can take now to help save the planet. I call her the Brené Brown of the environmental movement because she makes environmental action personal, doable, and joyful. Coming to terms with the impacts of climate change on our shared future and understanding we're at code red for humanity can make us all feel vulnerable and scared. But we have to embrace this vulnerability, experience all the emotions that come with it, and also celebrate the joy of climate action.

Heather provides advice without being preachy. We have all been talked down to enough, and that's not going to work for anyone. We have to abandon all-or-nothing thinking and perfectionism in the environmental movement, which is why I love the concept of a "one green thing" to ease eco-anxiety and shift the culture to a greener future. Everyone is welcome. And you can start today. Right now.

Since the landmark water contamination case in Hinkley, California, and the film that told the story and bears my name, I hear from people across the country, and I show up to help whenever and wherever I can. But as I've said time and time again in my books, speeches, TV appearances, and even in the ABC TV show *Rebel*—no one is coming to the rescue. We have to stand up for each other because Superman isn't coming. It's up to us.

We all have a role in climate action. *This book* will help you find how you can apply your talents in service to your family, community, and the climate movement. With determination and what I call "stick-to-itiveness," we can do it. We have the power.

The planet needs you. We all do.

ERIN BROCKOVICH
Consumer Advocate & Environmental Advocate
Author of *Superman's Not Coming* and *The Brockovich Report*

INTRODUCTION

"Mom and Dad, we are running out of time. I can't vote. You can't wait for us to clean up your mess and fix it. We need you to act now," my then fourteen-year-old daughter Cady pleaded.

It was September 2019. We were talking to Cady and our younger daughter Susan about the upcoming climate strike and student walkout inspired by young Swedish activist Greta Thunberg. The weather report called for heavy rain, so I offered to pick up Cady and drive her to the protest site after she left the high school. This parental gesture made perfect sense to me since she had to carry her trumpet and her freshman backpack, which weighed a ton. Besides, the protest starting point was a mile away.

Cady rolled her eyes and patiently explained to me, her environmental lawyer mother, that having a parent drive her to a climate walkout defeated its purpose. She said she was sick of all the praise for Gen Z (those born after 1997), that the planet was burning, and what were the Gen Xers and Boomers going to do about it? And then came her quiet tears. This response was the result of "eco-anxiety."

Ten years earlier, I'd heard about eco-anxiety from my colleagues Kevin Coyle and Dr. Lise Van Susteren of the National Wildlife Federation.[1] They wrote a paper about the mental health impacts of climate change, but as the mom of two young kids, dealing with the issue seemed far off. After my conversation with Cady, when I asked other Gen X moms about eco-anxiety, they immediately understood. Although we grew up on John Hughes films, were "latchkey" kids, and stressed about Cold War nuclear annihilation, Gen Z is different. One parent told me that her twenty-year-old son asked what the wildfires were like in Northern California when he was little; he had no idea that "fire season" is a recent phenomenon there. One friend's ten-year-old loves to draw Godzilla. He explained that the mythic creature symbolizes our relationship to nature, then matter-of-factly told her it's too late to save the planet. Teen climate leader Jamie Margolin, who suffers from clinical anxiety, told the *New York Times* in 2020 that climate change is like Beyoncé. She says there wasn't a time in her life that she didn't know about either.[2] And to be clear, eco-anxiety isn't something that only rich, privileged white folks experience. In poll after poll, Black, Indigenous and People of Color communities are more alarmed by climate change than other demographics and understand that communities of color are typically most impacted by climate disruption.[3]

Cady's tears at the dinner table marked the moment this issue of eco-anxiety and intergenerational action to protect the environment hit home for me. Even after 20 years of experience in the environmental field, if my kids felt alone in climate action, what was going on around other dinner tables? I could pick up Cady's backpack and trumpet, but how would I encourage more people to address the overwhelming problem of the climate crisis?

My dinner table conversation changed my professional direction. I knew right then and there that I had to do two things. First, I needed to understand the mental health impacts of the climate crisis. Second, I had to create a way for more people to "see themselves" in climate action. We needed these young people to know that they were not alone. I started my research, wrote this book, and created a nonprofit organization OneGreenThing.org.

As you'll learn in this book, I realized that a daily practice of sustainability, what a call a "one green thing" could ease eco-anxiety, give people a sense of agency in this overwhelming problem, bring more joy into their lives. And create the necessary culture shift to scale climate solutions. That's what this book is about.

But, the concept of how to create this shift—a "One Green Thing"—began years ago, way before I had kids. Think peak '90s: grunge music, jeans, and lots of flannel.

During my break from studying conservation biology at the University of Otago in New Zealand, I grabbed my overstuffed green Kelty backpack and hitched a ride to Lake Tekapo. As a twenty-one-year-old international student, I was there to view the scenery, not to visit the stone church at the edge of the lake. Nevertheless, I was drawn to it. I paused my yellow Walkman that was blaring a mix tape of Ani DiFranco, Digable Planets, and Smashing Pumpkins, removed my headphones, and sat in one of the wooden pews. Behind the altar a floor-to-ceiling window showcased the stunning, expansive aquamarine lake. My geology professor once told me that minerals in the glacial rock till reflected the sunlight, resulting in the impossibly blue color. The snowcapped mountains filled me with a sense of splendor and awe that pulsed through my body. It was exhilarating . . . and weird.

I stepped out of the church and sat by the lake in silence. I stared at the mountains, breathed in the air, took off my clunky hiking boots and socks, and stuck my bare feet in the earth. Stillness was rare for me, but this moment of connection with nature was powerful.

Inspired in part by Vice President Al Gore's *Earth in the Balance*, published during my second year of college, I wanted to learn more about climate change. Science has always intrigued me. I grew up in East Tennessee surrounded by the Great Smoky Mountains. I also knew that gorgeous, natural setting was plagued by cancer clusters, toxic nuclear waste, and strip mining. When I was twelve, my dad and I traveled from Tennessee to Yellowstone in a bright red sports car. The national parks and the beauty of the American West sparked my fascination with geology, ecology, and culture.

But the quiet contemplation at Lake Tekapo tied it together: my love for science, my passion for wild places, and my interest in policy. The striking views from the small church inspired me to look inward. What did I want to do with my life? Who did I want to be? I remember thinking that if I ever had grandchildren, I would want them to experience places like this, to savor the outdoors and the wonder that nature brings. I decided then and there to become an environmental lawyer to protect these remarkable landscapes for the next generation.

That day I wrote a letter to my parents and told them my revelation about my future career. The moment of contemplation also stuck. From then on, I tried to reflect on my relationship to the environment each day. I went to law school, practiced law for a while, joined political campaigns—including Al Gore's presidential run in 2000—worked in nonprofit environmental advocacy, and then moved on to Capitol Hill. I ran a nonprofit environmental health think tank in Washington, DC, and then led the nonprofit partner to Yellowstone National Park. Through all these roles one concept kept coming back to me: the One Green Thing. At Lake Tekapo I discovered that connecting to the environment wasn't a moment or a calling—it was a practice.

Let's face it: most of the research about toxic chemical pollution, water quality, our food, the rapidly changing climate, and species extinction is overwhelming and depressing. In 2020 alone we witnessed more than 42 million acres burn in Australia; so many Atlantic tropical storms and hurricanes that we ran out of names and resorted to the Greek alphabet; temperatures of 100 degrees Fahrenheit in the Arctic Circle; and destructive wildfires in California, Oregon, and Colorado that brought apocalyptic orange skies.[9] Scientists warn that we have six years before the climate crisis reaches a point of no return.[10] That's the inconvenient and unsettling truth.

It's little wonder people are experiencing "eco-anxiety," this intense emotional stress about climate change. A recent global survey of ten thousand students found that nearly half of young people say that eco-anxiety is interfering with their daily life.[11] Gen Z also understands that climate policy must center racial, economic, and social justice, known as "climate justice," as we build a new paradigm of a regenerative, hopeful future.[12]

We know that one person can't solve the climate crisis and that we need comprehensive policy and market action. But being intentional with how we live each day—from what we eat to how we connect to our spirit, our community, and the earth—will give us a shot at a greener, healthier future by changing the culture. We ***can*** save our sanity, our kids, and the future. Most of all, celebrating a daily practice of sustainability is uplifting and fun. That's where this book comes in.

UNDERSTAND ECO-ANXIETY & EMBRACE THE LAWS OF CHANGE

The Oxford English dictionary defined the term "eco-anxiety" in September 2021 as "unease or apprehension about current and future harm to the environment, human activity, and climate change."[13] Though a relatively new term, more and more people are experiencing actual trauma and also extreme worry about the future because of climate-fueled disasters. It's time to recognize the weight of inaction on climate to our future and our collective mental health, but also to pivot collectively and exercise agency over this global issue. Systems change only comes when individuals act.

From my twenty years of experience in Washington, DC, and the American West, I know the ins and outs of environmental science, law, and policy. Lobbying multiple presidential administrations on energy and climate policy, advising United States Senator Russ Feingold on environmental issues, and serving as a staffer for political campaigns means I understand something beyond climate science: I know we must have cultural change for policy to work. This is where your One Green Thing comes into play. A repeated, uplifting act can alter how we think and feel and eventually shift public opinion in favor of bold climate policy.

The most well-intentioned people often feel overwhelmed when they grasp the enormity of the climate crisis. But we can turn a sense of helplessness into a sense of accomplishment with small, consistent actions. Setting an intention each day to take a step—a One Green Thing—to care for the planet can help ease our anxiety about the future, push the culture toward climate solutions, and create a sense of joy.

These routine habits can shift our collective consciousness to support comprehensive climate solutions. This realization motivated me to write this book, which is a call to action to process your climate emotions, address your worries about the future, then identify your Service Superpower and create a daily sustainability practice to help protect the environment. To inspire others to get involved, I also started a nonprofit organization, OneGreenThing, which tackles the mental health impacts of climate change.

We'll delve more into eco-anxiety and the Laws of Change in the next chapter, but for now here's a brief overview:

- Eco-Anxiety: Defined as "the chronic fear of environmental doom," this term encompasses young people's worry about the future.
- The Eco-Anxiety Trifecta: Generalized anxiety, chronic loneliness, and a hyperawareness of the climate emergency—driven by social media—contribute to this phenomenon.
- Law of Simplicity & Consistency: Simple, consistent, daily actions are more likely to stick.
- Law of Identity: A habit must become part of who you are in order to take hold.
- Law of Amplification: Sharing our experiences makes change reverberate, resulting in impact and generating support for policy solutions.

DISCOVER YOUR SERVICE SUPERPOWER TO HELP SAVE THE PLANET

The Service Superpower Assessment I developed after twenty years in the field of environmental advocacy identifies which of the seven essential service types best suits your personality. According to the Law of Identity, this identity match means you are more likely to maintain your new habits of sustainability. The assessment also provides a powerful way for you to discover how to contribute to the movement. The challenge for most of us is the basic but tough question: *How do I take the first step?*

Once you've identified your Service Superpower, this book helps you pivot to adopt the One Green Thing mindset and create a daily intention of action.

Starting with a One Green Thing can be as easy as writer Anne Lamott's famous advice: "Go Outside. Look Up. The Secret of Life."[14] Of course, going outside to look up for five minutes isn't going to solve the climate emergency. However, that five minutes can restore you. It can inspire you. It can change your perspective. It can bring you hope. Even better, the action of going outside and observing can make you look forward to the next day's One Green Thing. Who knows? You might even discover your own version of Lamott's famous advice with a few of these small but compoundable actions.

HOW TO USE THIS BOOK

At its core this book and this daily practice is about compassion, community connection, and a hopeful, exciting vision for a sustainable and just future. The book is divided into three parts. Part 1 is a self-discovery guide for you to see how you can leverage your unique talents and get involved. It explains eco-anxiety and the Laws of Change, presents the Service Superpower Assessment, and includes a chapter dedicated to each Service Superpower profile, which contains

- an overview of attributes, strengths, and challenges;
- interviews with "Eco Heroes" who embody each profile;
- a list of One Green Things suitable for your Service Superpower; and
- a 21-Day Kickstarter Plan, the road map to start on your One Green Thing journey.

Then part 1 leads you through a visualization exercise, in which you'll envision 2030. This is the year when scientists think we'll need to have made significant action to curb climate change, or we'll face irreversible harm. I then ask you to imagine having a conversation in 2050 with a young person you're related to. You're their ancestor. What will they thank you for? What will they wish you had known? Next I outline ways you can apply your Service Superpower to positively impact our future, including an Eco-Action Plan and action tracker.

Part 2 addresses the seven areas to effect change so you can apply your Service Superpower in a daily practice of sustainability to ease eco-anxiety, a more in-depth discussion about how young people feel about the future we're leaving them, and includes journal prompts to help inspire action, reflect on your experiences, and track how you feel during the process.

Part 3 of the book is short and clear: change is up to you. This section provides resources for more detailed action as you incorporate a One Green Thing into your daily life.

You don't have to be a backpacker, travel to New Zealand

(although I highly recommend it!), or work on Capitol Hill to be part of the climate movement. We have a short period of time to come together as a society and make transformational changes to curb the impacts of this crisis.

Learning from my kid's eco-anxiety and their stress about the future transformed my research and my calling. The climate emergency is the biggest challenge of our time, and we all have a unique role to play. My mission is to help you find yours. Everyone is needed, and everyone is welcome. Come as you are. Start here. Start now. Listen to the young people you love. Let them know they are not alone in creating a greener, healthier, more just future. Discover and apply your Service Superpower. Let's be part of the solution, together.

Yours in partnership for a healthier,
greener, more equitable world,

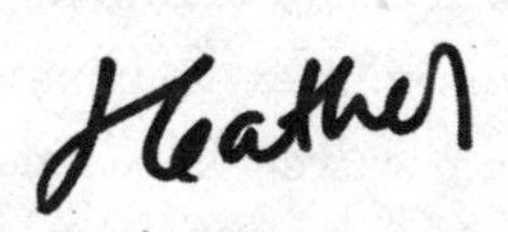

Part One

ECO-ANXIETY & THE LAWS OF CHANGE

Chapter 1

GET TO KNOW ECO-ANXIETY & THE LAWS OF CHANGE

Late one night during the pandemic, I told Cady to put down her phone and added that she wouldn't miss out on anything. It was an (online) school night. My anxious teen handed over her phone then remarked, "Mom, I'm not on social media because of FOMO [the fear of missing out]. My fear is the impermanence of human existence." I gulped. She'd seen in the news that her birthplace, Medford, Oregon, and an estimated 500,000 people in the state were on evacuation notice because of raging forest fires.

My daughter and I joked (or not) about a new acronym for Gen Z's constant screen use during the crises of 2020: it's FODO, the "fear of [humans] dying out." This is reality for Generation Z. They feel the sands through the hourglass, and not in a good way.

"ECO-ANXIETY" DEFINED: IT'S GLOBAL WARMING AND MORE

The term *eco-anxiety,* also called "climate anxiety," is a relatively new trend that many doctors and psychologists are witnessing. In 2017, the American Psychological Association recognized eco-anxiety as a "chronic fear of environmental doom."[4] Other terms include: climate anxiety, solastalgia, climate fear, climate grief, climate doom. The Climate Psychology Alliance and Climate Psychiatry Alliance formed to train mental health professionals to identify and treat eco-anxiety. A recent survey of child psychiatrists in the United

Kingdom discovered that 50 percent had clients who suffered from it. In a September 2021 international survey, one in four young people (ages sixteen to twenty-five) said that they likely won't have children because of their worry about the climate crisis.[5]

THE ECO-ANXIETY TRIFECTA: ANXIETY, LONELINESS, AND ENVIRONMENTAL STRESS

For Gen Z, "eco-anxiety" has three aspects. First, *children are suffering from generalized anxiety in greater numbers*. Each child experiences anxiety differently, but the statistics are alarming. The National Institutes of Health indicates that 30 percent of American teens suffer from anxiety. Rates of teen anxiety, depression, and suicide have dramatically increased since 2011.[6]

Second, *Gen Z is the loneliest generation*. More screen time and less in-person interaction mean a sense of isolation for them, even before the pandemic. In the 2018 Cigna Loneliness Index, Gen Z expressed "feeling like people around them are not really with them (69 percent), feeling shy (69 percent), and feeling like no one really knows them well (68 percent)." Our kids are lonelier than the elderly. In one survey, eight in ten Gen Zers experienced loneliness compared to five in ten Baby Boomers.

Third is what Cady and I termed FODO, the *hyperawareness of the climate crisis*. Gen Z is asking what the future will look like, where they will live, and what their experience will be like on this planet. In a 2020 survey by the US Conference on Mayors, 80 percent of Gen Z agrees that "climate change is a major threat to life on earth"; one in four have taken direct action on climate change, and by three to one, Gen Z believes "the climate crisis warrants bold action."[7] They know that we must act fast. The coronavirus pandemic cracked open the truth of the intersectionality of public health, systemic racism, the economy, and the environment. Now the concept of eco-anxiety encompasses a generalized anxiety about the future.

In today's culture we go to extraordinary lengths to help children avoid discomfort (like offering to pick them up in the rain and drive them to a climate protest as I recount in the Introduction) to ease our own anxiety about our children's pain. Yet recent research shows that kids with clinical anxiety have to be part of the solution in dealing with their stress. Fixing it for them doesn't help.[8]

Like my daughters, the majority of Gen Zers are worried about climate change. A friend told me that of course my kids have eco-anxiety—it's because they are *my* kids. I talk, write, and think about the climate crisis, so my kids would naturally be more aware of the issues. I encouraged her to ask her sons about it at dinner that night, and they said, "We think about climate change all the time, Mom." They just hadn't talked about it as a family.

It's time to address the mental health impacts of this crisis. It starts with intergenerational conversation, action, and a call to "think beyond our age," as I discuss in Chapter 13. If climate anxiety interferes with your daily life, seek a mental health professional. In many instances, action can help abate anxiety. But knowing where to begin your climate action journey isn't as easy as it seems.

More than twenty-five years ago, I climbed Mount Kilimanjaro while spending a summer studying international environmental law in Nairobi, Kenya. I boarded a bus to Arusha, Tanzania, then started my trek up the famous 19,341-foot mountain. I wasn't a climber. I was a policy geek, which is my nice way of saying that I wasn't physically fit (I could write killer legal memos though).

Kilimanjaro is a different kind of climb. It's basically a walk.

The guides told us to go slowly, to acclimate to the lack of oxygen at higher elevations. My classmates and I relished the journey as we enjoyed the views, took pictures, and laughed.

A group of ultrafit British women were making the climb with another tour company. They always asked us how long it took us to get to the next hut, which was our destination each evening after an all-day hike. I remember one hiker laughed at us and said, "I cannot believe how slow you are." We didn't know these women. And why did they care how long it was taking us? It's not as though our pace was impeding their progress.

Eventually we reached the top of the mountain, Uhuru Peak. On our way down we learned that none of the Brits had reached the summit. Altitude sickness overcame them. They'd rushed to each hut but didn't let their bodies adapt to the oxygen-starved air.

As my Southern momma would say, bless their hearts.

I remember thinking that some people assume environmentalists have a similar attitude to those British hikers: if you're not doing it our way, you're doing it wrong. Or people believe environmentalists think, *You don't belong here because you don't care enough, don't know enough, and don't understand the issues the way we do.* My colleagues in the environmental movement are nothing like this, but many people have an understandable fear of being judged for "doing sustainability" wrong.

Then there's the enormity of the climate crisis itself. Heat waves in the Pacific Northwest, tragic fires throughout the West, hurricanes and floods in the Southeast, melting glaciers, and rapid species extinction around the world. How do we even begin?

Because of climate change, the snows of Kilimanjaro are much different today than they were twenty-five years ago. It's true that we can't take it slow when dealing with this crisis. And the movement needs everyone—including *you*. If you are struggling with eco-anxiety and aren't sure where or how to start, this book aims to help you launch your journey. If you've been involved for a long time and are feeling burned out, I hope this book will restore you, invigorate you, and infuse even more creativity and inspiration into your work. Let's identify your unique talents and gifts and create an action plan to be part of the movement. Start now, but go slowly at first. Then your daily actions will help manage your worries and create the broad support for climate policy and market solutions we need.

THE LAWS OF CHANGE

When I was in law school in the late 1990s, we needed dimes and quarters to make copies at the law library. Yes, we made paper copies of cases and articles by xeroxing them from books. In the lobby of the grand law library, right in front of the librarians' desk, was a large sign that read: "We cannot make change." Can't make change? At a law school? At my beloved law school? Of course, it wasn't a message of the futility of working toward social change—the librarians simply meant they couldn't exchange dollar bills for coins to make copies. But the sign cracked me up every time I entered that beautiful space to study. After all, I went there to make change.

Unlike the law librarians at my alma mater, I want to make sure you know that *you* can make change.

The Laws of Change are the foundation of this book and the blueprint to discovering your Service Superpower and saving the planet. We each have to find our Service Superpower—our unique way of helping others. Everyone has a different skill set and personality for helping lead the necessary culture change through bold, positive action to support a healthier, greener, more equitable future.

Many argue that individual action doesn't matter in an issue as global and enormous as the climate crisis. They are wrong. Individual actions matter, but perhaps not in the way you think. You alone will not solve the climate crisis. Neither will I. But if you intentionally live a more sustainable life and connect with your community about your daily practice of One Green Thing, you can build momentum for *culture change* to shift policy. For example, Bret Jenks, president of the international conservation nonprofit RARE, recently told the *Atlantic*, "If 5% of Americans bought carbon offsets or changed other [carbon-intensive] behaviors, that would add up to a reduction of 600 million tons of carbon dioxide a year," and that shift would be "one of the top changes in terms of greenhouse-gas emissions in human history."[9]

Taking action can also help reduce anxiety about the future. The compounding power of daily rituals can not only transform your personal relationship to the planet but also create a ripple effect from your life to the lives of people you know. Simply put, the Law of Simplicity & Consistency, the Law of Identity, and the Law of Amplification work together to create culture change.

The Law of Simplicity & Consistency

"The only goal is to brush your teeth for two minutes in the morning. That's it. For a lot of my troubled, at-risk students, this simple action created a snowball effect. They got out of bed, got dressed, and even went to school. I made sure they didn't keep the water running while brushing. Thought you'd be proud of that," my retired guidance counselor mom told me in her soothing Southern accent.

I was both proud and inspired. Daily teeth brushing embodies the Law of Simplicity & Consistency, and that act became my unexpected metaphor for addressing the existential threat of climate change.

As previously stated, small actions alone will not solve the climate crisis and neither will the actions of one person. Only one hundred companies are responsible for 71 percent of greenhouse gas emissions since 1988. And only 8 percent of plastics are recycled.[10] We need substantial policy and market solutions, but individual action can create culture change so these comprehensive strategies work.

Tiny changes create momentum. You can find a ton of research on this phenomenon in leadership books, but my favorite resource is James Clear's *Atomic Habits*. He reminds us that "we often dismiss small changes because they don't seem to matter very much in the moment."[11] But if we create systems that allow for small changes, their impact compounds over time. Our daily actions make up our lives. If we focus on tiny, consistent actions, we can make time itself a powerful agent for transformation of our individual and collective experiences. Just like my guidance counselor mom telling her at-risk students to set a goal of brushing their teeth for two minutes each morning, micro steps snowball.

To make a habit repeatable, you need it to be obvious, attractive, easy, and fun.[12] Another way to think about creating a new habit is focusing on cue, action, and reward, as Charles Duhigg writes about in *The Power of Habit*. This means that something reminds you to take action (the "cue"), you take the action, and then your brain is rewarded for that action. Naming the reward is powerful enough to create a new habit.[13] In the context of One Green Thing, the feeling of positive contribution to the planet and your community can be the reward.

The Law of Identity

In addition to the cue and reward, we must embrace what I call the Law of Identity to make habits—such as a daily ritual of sustainability—stick. If you view the ritual as part of *who you are*, you're more likely to sustain it. For example, Clear points out that if you want to be known as someone who is physically fit, you'd ask yourself, *What would a healthy person do every day to achieve their goals?* Then you'd take small, daily actions—such as adding more steps to your day, skipping your morning venti latte in favor of a mug of green tea, and so forth—and keep checking in until you've reached your goal. As Clear communicates about the power of identity, "Decide the type of person you want to be. Prove it to yourself with small wins."[14]

With the Service Superpower Assessment, chapter 2 focuses on helping you align your identity with specific actions—your unique way to support a healthier, greener, more equitable future. Everyone has a different skill set and personality to help lead the necessary culture change for big, positive action. Through this book, you'll tap into what you likely already know about yourself and how you show up in service. Ask yourself who you want to be: How do you care for the important people in your life? Are you a fixer? A listener? The person who plans meals for someone who is sick? The person who sends a friend a good book or a poignant song when they are grieving?

When you're developing your daily practice of sustainability, remember that the power of your One Green Thing isn't merely doing the actions but aligning the activities with your identity. As American politician, author, and educator Shirley Chisholm once said, "Service is the rent we pay for the privilege of living on this earth."[15] How do you want to pay it forward?

The Law of Amplification

In 2017, Robert Kelly, an international relations expert on the conflict between North and South Korea, readied himself for a live BBC interview from his home. Shortly after the cameras started rolling, Robert's four-year-old daughter, Marion, bounced into his office to see what was going on. Her adorable glasses and her now-famous and unmistakable walk showed her exuberance about what her dad was doing. Then, with almost perfect comedic timing, a baby walker appeared in the doorway. Robert's infant son was ready to join the excitement. Moments later Robert's wife, Jung-a-Kim, darted in to pull the kids out of the office. As Robert smiled and chortled a "Pardon me" to the announcer, his wife ducked to avoid the camera angle and scurried the kids back through the doorway. As if scripted, she then athletically lunged to close the door behind her. To date, this forty-three-second video clip has forty-four million views on YouTube.[16] You've probably seen the video. I bet you're even giggling as you remember it.

What made this video go viral? Marion's walk was so full of childlike energy and curiosity, you could not help but smile. Robert told the *Today* show that Marion was in a "hippity-hoppity mood" as she jumped onto the TV screen.[17] Many parents could relate to what Robert and Jung-a-Kim must have felt: that sense that we're not in control of a situation even though we're supposed to be in charge. The joyful essence amplified the moment. It wasn't staged. It was real life, both vulnerable and funny.

When most of us think about the Internet, we focus on the downsides. Of course, there are legitimate concerns about digital addiction, nature deficit disorder, and all the negativity and shame that abound online—especially during election season. This viral clip, however, shows the powerful force of joy that can be amplified online, even if manufactured algorithms are pushing for more extreme content and negative engagement.

Peer-reviewed psychological studies show that joy is contagious. This maxim is the essence of the Law of Amplification, which means that positive actions lead to more positive actions, and that sharing joy inspires others. The Law of Amplification is important to movement building. A 2008 study by Harvard and the University of California showed that happy people amplify their happiness to spouses, neighbors, and friends. Someone who experiences happiness and shares it can increase a spouse's happiness by 8 percent and a next-door neighbor's by 34 percent. Even better, that positive exchange can extend to three degrees of friends, reaching outside of the happy person's network.[18] In another study that reviewed thousands of *New York Times* articles, two University of Pennsylvania professors found that readers share positive stories more than negative ones. Additional research shows that reading others' positive Facebook posts can increase happiness by 64 percent.[19] The closer the real-life relationship of the poster and the viewer, the greater the shared happiness experienced.

Sharing makes a difference, and when you share your joyful experiences, the impact of the repeated daily action amplifies. Recruiting friends and family to try out One Green Thing creates an opportunity to fight the feeling of being overwhelmed and to hope for a greener world. As more join and our actions multiply, you can help the culture shift.[20]

In chapter 11, we'll discuss the tools for you to address eco-anxiety and support your daily practice of sustainability, including developing a personalized Eco-Action Plan, using the Eco-Impact Top Ten, and recording the mental health aspects of your practice with the Joy Tracker.

21-DAY KICKSTARTER PLAN

Once you identify your Service Superpower, you can read a detailed chapter that outlines your strengths and attributes, areas for development, and suggestions for how to create your One Green Thing. Most of us have heard that it takes twenty-one days to build a habit. That's why I created these plans. Imagine my surprise when I learned from behavioral scientists that it actually can take anywhere from eighteen days to almost a year to develop a new habit.[21]

Even so, I think you'll find that creating a daily practice of sustainability will bring enough joy to your life and help manage climate anxiety that it makes sense to give it a go for twenty-one days. *If it takes longer than twenty-one days for your daily practice of sustainability to stick, that's okay.* There's no guilt or judgment here, no goal of perfection. We'll start with how each of us has a role to play to make the movement accessible, positive, forward-looking, and even fun.

CHAPTER TAKEAWAYS

- Eco-anxiety also known as "climate anxiety" is the chronic fear of environmental doom. Young people are suffering and worried about their uncertain future because of climate change.

- To create systemic change, we all need to get involved. We each have unique Service Superpowers to apply to climate action.

- There are three Laws of Change:

    - The *Law of Simplicity & Consistency* reflects the compounding impact of small changes made over time.

- The *Law of Identity* means that you are more likely to stick with a new behavior if you align it with *who* you are.
- The *Law of Amplification* translates into sharing your joy to inspire others.
- Together these three laws—supported by the tools in this book—drive culture change for big climate policy and market solutions.

- Each Service Superpower profile chapter has a 21-Day Kickstarter Plan to get you started on your One Green Thing journey.
- Chapter 11 contains a detailed list of tools for you to manage eco-anxiety, handle the emotions that come with it, and move into action by creating your personalized Eco-Action Plan, measuring impact with the Eco-Impact Top Ten, and tracking your mental health with the Joy Tracker.

JOURNAL PROMPTS

- What inspired you to pick up this book? What do you want to learn more about?
- Do you know a GenZer with climate anxiety? If not, have you experienced this type of anxiety? Have you talked to a young person you love about climate change? How did it go?
- Have you ever been hesitant to start on your climate journey? If so, why? For instance, maybe you didn't know where to start, or you were worried about being judged.
- Reread the section on the Laws of Change. Have you ever experienced the Law of Simplicity & Consistency? Or the Law of Identity? When has the Law of Amplification impacted your life? For example, perhaps a friend shared a social media message or inspiring story that made you think differently or take action.
- Think about your daily routine. What are you already doing that might be considered a One Green Thing?

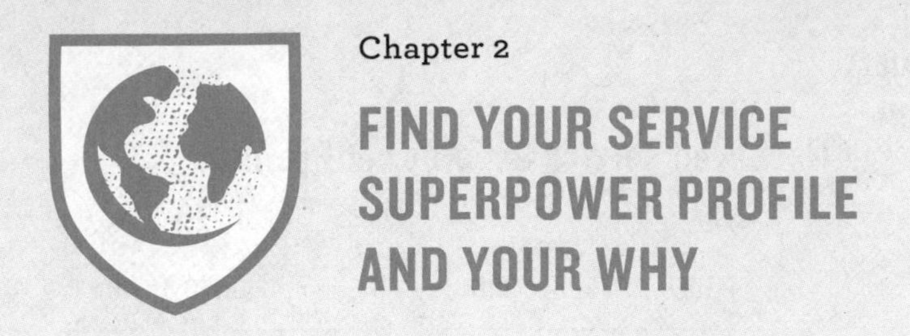

Chapter 2

FIND YOUR SERVICE SUPERPOWER PROFILE AND YOUR WHY

I developed the Service Superpower Assessment based on twenty years in the field of environmental advocacy. The goal is to determine which of the seven essential service types best suits your personality. This information will help you discover sustainability habits that are more likely to stick, according to the Law of Identity. Please visit www.onegreenthing.org/assessment to take the assessment, which takes approximately five minutes.

INTERPRETING YOUR SERVICE SUPERPOWER ASSESSMENT RESULTS

The following table shares the icon associated with each profile, along with a brief description and associated actions.

COMPOSITE PROFILES

The Service Superpower Assessment also recognizes that some personalities might gravitate one direction but have a strong inclination toward another profile type, which is true if you have a nearly equal number of responses for two profiles. Here are common combinations of Service Superpowers:

The Seven Service Superpower Profiles

PROFILE	PROFILE DESCRIPTIONS	ALIGNED ACTIONS
ADVENTURER	• Experiences the outdoors through challenges and hands-on learning • Embraces the physicality of time outside and gives family and friends opportunities to stretch their minds and bodies through nature	• Organizing a family hike on a local trail • Going kayaking with friends • Speaking at a local school about the benefits of time outside
BEACON	• Stands up to power and speaks the truth • Possesses a strong sense of justice and works to empower others • Serves as a skillful organizer and is often on the podium or behind the megaphone • Urges friends to stand up, speak out, and take action • Celebrates collective action and movement building	• Recruiting friends to participate in a beach cleanup • Testifying before the county commission on the need for a sustainability plan • Writing a letter to a member of Congress about why they need to support clean energy policy
INFLUENCER	• Follows the latest trends on sustainability, environmental action, and justice • Shares information and events with friends, introduces friends to one another, and motivates them to take action	• Sharing the latest articles on environmental issues • Encouraging friends to attend a climate policy lecture • Telling friends about favorite sustainable brands and urging them to make the switch

PHILANTHROPIST	• Excitedly serves by volunteering time, financial resources, and connections to important causes • Asks friends to donate, give time, and attend environmental events and experiences • Connects like-minded people together	• Volunteering at a local nature center • Organizing a fundraiser for a national climate group • Connecting like-minded friends to attend a lecture or event for your favorite conservation nonprofit
SAGE	• Focuses on the spiritual connection to nature • Is thoughtful, contemplative, and often quiet • Encourages friends and family to look for the deeper meaning of religious texts • Considers environmental protection a moral and ethical issue	• Taking a quiet walk in the woods • Reading a favorite psalm or spiritual text outdoors • Google the Climate Mental Health Network's "Climate Emotions Wheel" and share with your community
SPARK	• Ignites the movement and is the first to join an event • Loves to support others' passions and is always ready to cheer for friends and family	• Saying yes when asked to attend a climate lecture • Going to a friend's fundraising event for a wildlife group • Tagging along with friends to volunteer for a trail cleanup
WONK	• Loves science, policy, data, graphs, and charts • Enjoys discussing new technologies and solutions and the latest studies • Encourages others to learn more about the world around them	• Sharing favorite documentaries with friends • Giving a talk about climate solutions at a local community center • Writing an op-ed in the local paper about drinking-water pollution

- **The Adventurer-Sage** not only loves exploring but also views time outside as a deeply spiritual practice. It restores this composite, who expresses climate action through compassionate relationships with wildlife and nature.
- **The Sage-Philanthropist** is focused on finding common ground. Known as a peacemaker, this personality wants to use a strong sense of justice and moral duty, along with spirituality, to create space for mediation of conflict and for negotiated solutions. This composite profile shines in participatory decision-making and tough conversations about highly charged emotional and spiritual issues.
- **The Philanthropist-Wonk** composite often occurs in people who are in the environmental policy field. Sometimes they're scientists or policy wonks by training but also advocates focused on bringing more people and resources into the climate movement. This important combo can make complex social issues easier to understand and inspires people into action. (This is my profile type, if you haven't already guessed.)
- **The Spark-Philanthropist**, a common composite, is best described as an enthusiastic, action-oriented cheerleader. This profile reinforces that contributing to causes, sharing information and resources, and recruiting others to get involved all go hand-in-hand.
- **The Wonk-Spark Profile** is the operator. They mobilize the movement with a willingness to raise their hand, get involved, and translate ideas into action. This composite is highly analytical and organized, and it has a strong network of collaborators, colleagues, and community.

Your Service Superpower profile will reinforce your role in climate action through the Law of Identity. It will guide you on how to take this enormous, overwhelming issue and create your unique path to being part of the solution.

A WORD ABOUT SERVICE

When I talk about "service" in this book, I want to be clear that no individual can swoop in to save the day because they "know best." The service I'm referring to is how to help people and communities by giving back, restoring, regenerating, and supporting one another. Our only option as a species is to work together to save the planet. It will take all of us, united in service, to create a livable, thriving planet.

A WORD ABOUT EQUITY

Throughout this book, we'll address the issue of "equity," which is essential once you learn more about how BIPOC (Black, Indigenous, People of Color) communities suffer greater environmental harm from pollution. My definition comes from *Black's Law Dictionary*, which states that "environmental equity" is:

> Equitable sharing of environmental impacts by a community. Environmental policies and laws strive to ensure that no one group or community bears a larger, unfair share of harmful effects from pollution or environmental hazards.

> Economic or political clout is unnecessary by the development, implementation, and enforcement of these laws.[1]

We must also include Black, Indigenous, and People of Color and underresourced communities in development, enforcement, and implementation of these policies and actively invest in these communities.

A WORD ABOUT THE FUTURE

In this book I'll ask you to imagine the best future possible for the next generation. To stop global warming and to save the planet, we must be working toward something bold and exciting, with determination equal to stopping the old-fashioned, polluting ways of the past. I'll use terms like "regenerative," "circular," "sustainable," and "green." Here's what I mean: a future where we create systems, products, buildings, policies, and interactions that innovate, reuse, recycle, repair, and replenish the communities we live in and that the earth depends on. Think green buildings, rooftop gardens, high-speed trains, vibrant parks and open spaces, and clean energy, air, and water for all. I'll ask you to describe your green vision of the future in chapter 10.

CHAPTER TAKEAWAYS

- We each have a unique Service Superpower to apply to climate action. Learning your strengths, then creating a daily practice, is key to easing eco-anxiety and creating the necessary green culture shift.
- The Seven Service Superpower profiles are: Adventurer, Beacon, Influencer, Philanthropist, Sage, Spark, and Wonk. Composite profiles acknowledge the many different combinations of service profiles.
- Matching your Service Superpower to your daily practice, the Law of Consistency & Simplicity, reinforces the Law of Identity, which means you'll be more likely to take action.

For additional resources, visit www.onegreenthing.org/assessment.

JOURNAL PROMPTS

- Reflect on your Service Superpower Assessment results. Does the answer feel right to you? Have you embodied this profile in past service?
- Think about the activities you find the most meaningful when you're giving back. How could these types of activities be helpful in the climate movement? How might they help manage your climate anxiety?
- How does service make you feel? What areas of creating a brighter, greener future are you most passionate about?
- Consider your "why" to help protect the environment. Maybe you had a specific experience from your early life that instilled the desire in you. Or perhaps you're concerned about a specific aspect of climate change, or about the future your children, grandchildren, or loved ones will inherit.

Chapter 3

THE ADVENTURER

THE ADVENTURER PHILOSOPHY:
"Experiences, not things."

ADVENTURER ATTRIBUTES AND STRENGTHS

Attributes

The Adventurer is a tactile learner and lives for outdoor physical challenges and travel. This Service Superpower profile feels at home in the great outdoors and has a special relationship with animals and wildlife. Adventurers have a desire for hands-on learning and may have a hard time sitting still. They are highly creative and frequently excel in art or music. Pursuits like photography, art, climbing, kayaking, hiking, and cycling appeal to them. They love trying new foods, learning about different cultures, and expanding their understanding of how we interact with the environment. The Adventurer helps make the climate crisis real by challenging others to experience the outdoors through all five senses.

Strengths

Strengths of the Adventurer include a strong work ethic and stamina, a willingness to take risks, the ability to innovate, and the capacity to thrive outside their comfort zone. Given their creative tendencies and affinity for risk-taking, Adventurers can be disruptors and entrepreneurs. They're gifted educators since they can understand issues and experiences through all of their senses. Adventurers are usually great at logistics and highly organized in their own special way.

Grounded in physical space, Adventurers tend to have specific natural areas they hold dear, even though they love to fill up their passport with stamps. However, Adventurers don't have to travel. Some demonstrate their service personality through reading, research, and technology, creating immersive virtual experiences to help others feel empathy, see different perspectives, and take action. Because of their keen interest in culture and geography, Adventurers are more likely to have witnessed climate disruption firsthand. They therefore can be effective spokespeople for the need for change, as they recount their personal experiences with the climate crisis.

Adventurers are tough. They're comfortable with adversity and have overcome physical and mental challenges. With their strong sense of self, they are exceptionally good at managing crises or unanticipated situations. Quick on their feet and ready at a moment's notice, Adventurers use their spontaneity to urge others out of a rut or a fixed mindset. They inspire others to embrace change and to grow.

In service, Adventurers are challengers and encouragers. They have a surprising ability to identify others' oft-obscured talents. They can bring out the best in a team by encouraging the members to try new things, take on unfamiliar experiences, and gain a distinct perspective. These Adventurer traits are especially potent in service leadership because unexplored ways of thinking and problem-solving can help unravel complex societal and scientific issues.

ADVENTURER CHALLENGES

This profile's three goals for further development are to increase focus, to connect with others, and to set realistic goals.

Increase focus. The Adventurer mind is frequently someplace far away rather than in the present moment. The wanderlust that energizes their creative flow can also result in daydreaming, which means they miss critical details like deadlines or project needs—especially if the information is communicated aurally and not written down. Even though they're highly organized, most of this strength comes from managing their own expectations and own journey. This profile needs to lean into the Sage and the Wonk to be fully present and stay focused when serving others.

Connect with others. The Adventurer's constant commitment to action and striving for excellence means that they frequently end up on a solo journey. This profile could also benefit from leaning into the Spark. Adventurers can sometimes unintentionally spend a lot of time in solitude and need to "get out of their own head." To stay aligned with their personal service goals and remain linked to the larger cause, it's important for Adventurers to schedule time to connect with others.

Set realistic goals. Because Adventurers generally operate at a high level of innovation and creativity, they can be too lofty in their vision for change. Sometimes they need to break down their tasks and start with a more achievable goal. Reaching out to friends and teammates can help Adventurers set ambitious but realistic milestones on their journey to enact change.

ADVENTURER OVERVIEW

Traits: *creativity, boldness, agility, swiftness, determination*

Strengths: *hands-on, risk-taker, collaborative, strong, perceptive*

Challenges: *distracted, isolated, idealistic*

STRATEGIES FOR ADVENTURERS TO RESTORE AND REPLENISH

Adventurers are hardworking; in service they're organizing, leading, and educating, which can sap their energy. Here are ways for Adventurers to manage their fatigue from climate action and avoid burnout:

Do a barefoot grounding exercise. This helps the central nervous system relax and creates a sense of calm and connection. It may sound a bit new-agey, but standing with bare feet in the grass, dirt, or sand can help still the mind. Some researchers theorize that the electric connection between the ground and the feet can help create a positive immune response. Research has shown positive cardiovascular effects and deep relaxation from this exercise.[1] Taking ten minutes to plant their feet into the dirt, grass, or sand can make the Adventurer feel more grounded and present.

Spend time with friends. Even though the Adventurer is constantly on the go, they spend a lot of time alone, either planning travels and projects, conducting activities for friends and family, or bearing witness to climate disruption. Scheduling downtime with friends, such as a book club, a dinner, or a gentle walk outside, can help the Adventurer gain focus and restoration.

Channel creativity in unexpected ways. Adventurers are makers at heart. Trying something completely different, like taking a dance class or writing poetry, can help the Adventurer recharge. Mix it up by learning how to make jam or how to cross-stitch. An outside-of-the box project can trigger a unique perspective shift, thereby replenishing the Adventurer's energy and creative juices.

Adventurer Eco-Hero Profiles

Kristal Ambrose, PhD

Marine Scientist, Founder of the Bahamas Plastic Movement, and Goldman Environmental Prize Winner

Kristal Ambrose knew she wanted to study marine science at a young age. Growing up in the Bahamas, she spent her time studying coral reefs, scuba diving, conducting reef fish surveys, and providing general field support. That changed in 2012, when she heard Marcus Eriksen of the 5 Gyres Institute speak about an oil spill in the ocean—but the oil was a sea of plastic. Eriksen invited Kristal to sail to the Western Garbage Patch in the Pacific Ocean.

"Before I knew it, I was sailing across the Pacific Ocean to study garbage. I was the only Black person on the ship. I was in the sea of environmental conservation in a predominantly white culture." On her turn for cabin duty, Kristal discovered that she was the only passenger with a plastic reusable water bottle instead of a metal one. She realized she stuck out too. The lone plastic water bottle next to all the metal ones became a metaphor for her experience in marine science and conservation.

"There were no islands in the distance, no other boats, no airplanes flying overhead. It was just us, wildlife, and waste." The experience reminded her that "over time, you realize what matters. It's nature and divine-human connection."

That trip sent Kristal on a different adventure, as she wanted to learn more. "I encouraged students to help me track how plastic was moving over space and time in the Bahamas." She said that she'd joke, "This is the Bahamas Plastic Movement, guys." And about a year later she decided to start the nonprofit. "I kept envisioning a mass of locals walking with me to create a nation free of plastic pollution."

Now, eight years after founding the Bahamas Plastic Movement, Kristal recently completed her PhD in marine science in Malmo, Sweden, and she received the Goldman Prize, known as the "Nobel Prize for the Environment" for her exceptional work. "I think I won it not for what I did, but more for how I did it. That how is through empowerment and through joy."

Kristal understands that the issue of plastic pollution in the ocean can be depressing and that it's easy to get burned out. "But the ocean is still there. There's still hope out there. It's not about me. I'm just ordained to this work. It's about the children who look forward to this work every year. It's not just about plastic. Plastic is a catalyst for empowerment."

Having won global recognition reminds her to keep moving, to keep doing the work. Kristal declared that her Goldman Prize can encourage other Bahamian children: "If I can do it, they know they can. Life is limitless."

Chris Hill

Lawyer, Angler, and Activist

Eco Hero Christine Hill, known as "Chris," never thought of herself as an outdoorsy person. Then one summer her parents enrolled her in an outdoor camp session they'd won in a school fundraiser. Instead of being miserable, Chris loved rock climbing, hiking, and camping. "I fell in love with nature, with rock climbing, with the people. And then I asked if I could go the full summer. I went back for six summers and ended up becoming an instructor."

Chris also learned about leadership and nature. "This outdoor camp was the most pivotal moment in my life." It helped her realize "what being in nature means to me, how it rejuvenates my soul, my mind, how it makes me feel good. If I want to spend time in nature and want my kids and grandkids to have this experience, what do I need to do to protect it?"

A star student, Chris graduated high school a year early and went to college at Appalachian State University where she was drawn to the broadcast journalism school and the nearby stellar rock climbing. While studying abroad in Costa Rica, Chris worked with local lawyers and organizers to safeguard a small village from a planned gold mine. This sparked her interest in law school. She shifted her focus from broadcasting to law and lobbying and enrolled at Vermont Law School.

Chris knew she didn't want to be in a courtroom, but she wanted to effect change through federal policy that centered on communities' needs. Right after law school she worked with a nonprofit and witnessed mountaintop coal mining removal firsthand. "My first thought was, *This can't be happening in the United States*. I knew that the voices of these communities needed to be front and center in talking to legislators and the media." Now at the Sierra Club, Chris serves as its Chief Conservation Officer to protect the country's wild places in order to combat the climate crisis and ensure equitable access to the outdoors for all people.

Despite being the subject of the award-winning Banff Film Festival documentary *Where I Belong*, which is about creating outdoor spaces for BIPOC communities, Chris describes her leadership style as "behind the scenes." She aims to "lift up other people, push others forward, connect people, and especially lift up the next generation." What gives her hope is her belief in our shared humanity. "There are so many passionate people in this world, in our communities, that may or may not have tapped into what it means to be stewards of the environment. But with just a little information they get it."

Her advice is to begin where you are, and her favorite One Green Thing is talking to friends about nature, climate action, and environmental health. "Talking to your small circle of friends and family members makes a difference. Don't underestimate the power of social media and sharing your actions."

Asher Jay

National Geographic Explorer and Tech Entrepreneur

From Times Square to the TED stage, tech entrepreneur Asher's artwork inspires people across the planet to respect and protect wildlife. A world traveler, photographer, and artist, Asher was named a NatGeo Explorer by National Geographic.

Asher calls herself a "creative conservationist" who loved environmental science as a kid. "I read David Attenborough as a first or second grader, with a dictionary by my side. I was obsessed with the natural world" and "highly sensitive to global things, particularly the plight of the earth." One day she was watching a program about chimpanzees being experimented on and cried so hard that her parents intervened. "No more watching TV, no more books about wildlife," declared her father.

"I was way too emotional and consumed by the harm to the earth, from a place of victimhood and sorrow," she says. Asher had to find a different way. She was modeling at the time, and decided to go into fashion and studied at Parsons.

Then the 2010 BP oil spill happened. When she saw dolphins dying en masse and other wildlife covered in thick oil, environmental activism pulled her back. She recalled that the cause of the spill was an improperly installed drilling rig. "To think that a few bolts could result in the destruction of so much life broke my whole reality apart." When BP started spraying dispersants to break up the oil, the efforts to control the damage made the impact worse. "I knew I had to be the change. In 2009, I lost my dad. He had agency and chose his course of treatment. These dolphins, though, did not. Twenty thousand marine mammals dying was unacceptable. Every cell in my body was telling me I had to do something."

Asher channeled her pain into action. "I chased environmental thought leaders I wanted to meet and hopped in cabs to the airport with them to learn more. I wanted answers." Soon she was on TV talking about the BP spill and plunged headfirst into conservation, art, and the climate movement.

What gives Asher hope is her transformation from anger to acceptance and a firm belief that it'll be okay. Her advice is to start with connection. "Find a moment to consider a living thing outside yourself. Spend time with it. It could be a grasshopper that you notice or an inside house plant," stated Asher. "It's not a commodity. If you make that genuine connection outside of yourself, you can create space to show up more."

Her start-up aims to create radical transparency for consumers in transactions. Through a blockchain-based app, consumers can verify that brands are sustainably producing their goods. "My team is working 24–7, and we remind each other how much good we can create. That's what keeps us going."

WHAT NEXT?

Now that you've read the strengths, attributes, areas for development, ways to avoid burnout, and the Eco-Hero profiles, does the

Adventurer Service Superpower resonate with you? If not, explore other Service Superpower profiles. If yes, make a list of your unique skills in service as an Adventurer.

My strengths as an Adventurer are:

1. ____________________
2. ____________________
3. ____________________

Three things the climate movement could do a better job of are:

1. ____________________
2. ____________________
3. ____________________

After reflecting on my interests and talents, here are three ways I might be able to help:

1. ____________________
2. ____________________
3. ____________________

The 72-Hour Check Back

In the next 72 hours, tell someone significant in your life that you're embarking on a daily practice of sustainability, that you'd like their support, and that you'd like them to join you on your next adventure. Write that information below:

The person I told about my daily practice of sustainability is ____________________.

I invited them on my next adventure to ____________
(location) on ________________ (date).
Tomorrow my One Green Thing will be __________________.

ADVENTURER ONE GREEN THING EXAMPLES

Micro Impact

- Go birding with a friend to relax.
- Bike instead of driving to meet a friend.
- Invite friends from different generations for a hike or walk to your local park.

Macro Impact

- Lobby Congress to restore federal funding for river restoration.
- Visit a local community organization and share your personal experience witnessing climate disruption.

To start on your journey, check out this 21-Day Kickstarter Plan for the Adventurer to inspire your daily One Green Thing. As you apply the Laws of Change, you'll see that the Laws of Simplicity & Consistency, Identity, and Amplification mean that your small, daily actions compound to ease eco-anxiety, shift the culture, and promote comprehensive solutions to the climate crisis. You can see that the Kickstarter Plan aligns with part 2, "The Seven Areas to Effect Change."

Feel free to either review the other profiles or skip ahead to chapter 10 for a visualization exercise and chapter 11 to check out the tools, including the Eco-Action Plan, Eco-Impact Top Ten, and Joy Tracker. Then flip to part 2 and put your Service Superpower into action.

THE ADVENTURER

21-DAY KICKSTARTER PLAN

	SUNDAY *know you can't go it alone*	MONDAY *think beyond your age*	TUESDAY *see energy in a new light*
WEEK 1	Try out a yoga class online or in your community. Look for a variety of pace and skill levels. After attending, reflect on the connection of mind and body.	Share personal stories from times you have witnessed the impacts of climate change. Tell a person who is not in your generation or post on social media.	Buy carbon offsets for your next trip.
WEEK 2	Call a friend and take a walk or hike in a local park. Talk about what you see, why you care about the public space, and find out how you can help.	Go on a walk or hike with a neighbor or family member in a favorite local park or nature center. Take a field guide or download an app to identify what you see and hear.	Take ten minutes to find out whether your city or town has a sustainability director. Send an email asking to talk with them about the plan.
WEEK 3	Find out if there's a community garden near you. Sign up to volunteer and introduce yourself to others who are volunteering.	Call a mentor, older family member, or friend. Ask them about their thoughts on climate change and whether it's affected places they love.	Call your energy provider and ask about their renewable portfolio. Tell them that clean energy is important to you. Consider switching to clean power.

WEDNESDAY *understand that you are what you eat*	THURSDAY *protect the source*	FRIDAY *vote with your wallet*	SATURDAY *love your mother (earth)*
Make a commitment to eat locally grown or regionally native foods whenever planning your next adventure.	Google "citizen science" opportunities in your community and see if you can contribute to a water-quality or stream-monitoring research study for your next outing.	Make a commitment to thrift your outdoor gear, barter, or trade gear with a friend instead of making a new purchase for your next adventure.	Go on a walk with a neighbor or family member in a favorite local park or nature center. Take a field guide or download an app to identify what you see and hear.
Plan a trip to your local farmer's market. Try out a new vegetable you've never tried. Ask a farmer at the farmstand for more information about what they do, who they are.	Go to your kitchen. Write down all the plastic you see — from food packaging containers to detergent. Identify three things you can do to limit your plastic consumption.	Commit to buying nothing today or tomorrow.	Put your phone down. Go outside. Look up. Take ten minutes to sketch what you see, hear, and feel.
Try a plant-based meal from your favorite geographic area to explore for family and friends.	Research Wild & Scenic Rivers in your state. Plan a trip or take five minutes to write to your legislators about how important these waterways are to you.	Take a look at your closet. Declutter and designate items to thrift or give away.	Plan a day to volunteer at your favorite local public space or make a donation to its environmental education program.

Chapter 4

THE BEACON

THE BEACON PHILOSOPHY:
"Be the light; be the change."

BEACON ATTRIBUTES AND STRENGTHS

Attributes

The Beacon stands up to power and tells the truth. They're the attendees who will raise their hand in a town meeting and ask the tough questions when everyone else is silent. Or they'll be the ones standing in front of city hall at a podium or with a microphone in hand. People with this Service Superpower want to inspire, protect, and empower others. Beacons can be found in the ranks of entrepreneurs, activists, and organizers.

Strengths

Strengths of the Beacon include a willingness to speak out, fearlessness in having difficult conversations, an interest in trying creative strategies to familiar problems, an ability to articulate a clear, bold vision for the future, and an unwavering commitment to the underdog. Despite their focus on social justice and change, Beacons show up in service in various industries. From teachers to scientists, Beacons fit every job category.

Beacons are visionaries who approach problems with a strong sense of right and wrong. Their clear conviction and vision translate into an electrifying opportunity for others to be inspired by action. The Beacon sees the way forward, even if the path takes people to unexpected places. We need Beacons in the climate movement because we need advocates who keep telling business leaders and our elected representatives that we expect action on clean energy, sustainable design, green jobs, and climate education for the next generation.

Beacon energy means this profile transforms the feeling of a room from ordinary to a sense of possibility and excitement. Beacons get the ball rolling. They help people get off the couch, have a tough conversation, attend a rally, or respectfully disagree when someone says the climate crisis is overblown.

Their courage means that they're the first to take action, many times alone, and they're the vanguard of movements. The ability to stand their ground and not back down means that the Beacon can take the slings and arrows of being a driver for change.

BEACON CHALLENGES

Three big areas for Beacons to work on are to develop self-compassion, to cultivate patience, and to practice mindfulness.

Develop self-compassion. Beacons have high expectations for rapid change because they know the needs are great and the situation is dire. When things aren't progressing as they envision, Beacons can be tough on themselves. Also, because their empathy is high, they acutely feel others' pain. To counteract these tendencies, Beacons need to reach toward moments of joy and not be so serious all the time. This need to relax correlates to self-compassion.

Cultivate patience. Because this Service Superpower profile feels the urgency for change, patience isn't necessarily a strength. They understand where the movement and future can lead if everything lines up, so they want bold action, now, as well as swift progress. Beacons quickly process information and can connect the dots, and it's easy for them to get frustrated when others don't immediately grasp the issues or fully understand the need for change. Beacons need to take a deep breath, slow down, and walk their team through the issue and action steps. Working on active listening is essential for Beacons because they're usually "out front" in service and thus more accustomed to being listened to instead of being the listener. Sages and Wonks balance out Beacons, because those two profiles ground Beacons in the here and now and make them spell out the steps to solve the problem.

Practice mindfulness. *Mindfulness* is a buzzword these days, but there's a reason we all keep talking about it. Many of us—not just Beacons—are focused on the "busyness" of everyday life and forget to stop and look and be present. Phones have become such a pervasive distraction and disruptor of one-on-one communication that our brains have become accustomed to constant input. This leaves Service Superpower profiles like the Beacon without much time for self-reflection. Deep breathing, prayer, and meditation can help the Beacon be fully present with others and not so focused on the future, the next action, or the upcoming rally. The Beacon will benefit from a gratitude practice as well. Writing down five things you're grateful for each day has been shown to improve mental health.[1]

BEACON OVERVIEW

Traits: *futuristic optimism, forthrightness, curiosity, street smarts, courage*

Strengths: *visionary, clear, steadfast, righteous, fearless*

Challenges: *self-critical, impatient, rushed*

STRATEGIES FOR BEACONS TO RESTORE AND REPLENISH

Beacons will benefit by spending time in nature and leaning into their inner Sage for solitude and quiet. This Service Superpower profile tends to be go, go, go all the time. Their focus on action, their felt urgency of the need for change, and their deep empathy for the pain around them prevents them from taking downtime. Of course, nature is the best healer, so I'll always recommend that every Service Superpower take time outside—even a short walk to a local park—to restore themselves.

Three strategies for the Beacon to explore to restore and replenish include:

Complete a digital detox. A forty-eight-hour phone-down, tablet-down experience will help restore the Beacon. Because of their focus on activism, this break prevents them from using social media to talk about their next campaign or researching the latest legislative proposal on the issues they care about. The Beacon needs to use this time to embrace the quiet, even go outside, and think, feel, and experience the moment as it unfolds.

Practice meditation or prayer. A thirty-minute practice of prayer or meditation to clear the mind and focus on breathing is restorative for the Beacon. Visit onegreenthing.org/bookresources to access a guided meditation for relaxation.

Read fiction. This suggestion may seem to be a strange way for the Beacon to replenish, but since this Service Superpower profile is all about business and action, the act of reading a good story is revolutionary for this type. This activity helps the Beacon's brain turn off for a few minutes and immerse in a different flow of creativity, joy, and art appreciation.

Beacon Eco-Hero Profiles

Mustafa Santiago Ali

Vice President of Environmental Justice, Climate, and Community Resilience, National Wildlife Federation

"I got involved in environmental justice from my early days of listening and learning. My grandfather was involved with the union movement and civil rights in the coal mines of Appalachia," reminisced Mustafa Santiago Ali, a vice president of the National Wildlife Federation. "My dad focused on civil rights as well, so I was watching and thinking critically about solutions."

A Beacon Eco Hero, Mustafa is a regular expert in national print and television media, including MSNBC, CNN, and VICE. He currently cohosts a live radio show and podcast, *Think 100%: The Coolest Show on Climate Change*, with singer and actress Antonique Smith and civil rights icon Reverend Lennox Yearwood. His unyielding compassion energizes others to stand up for a greener, more equitable future.

As a child, Mustafa remembers seeing "good, hardworking folks exposed to toxic chemicals that were making them sick. I've also seen the beautiful nature around me destroyed by strip mining and its devastating impacts on community health in West Virginia. At sixteen, I knew this work was my calling."

Mustafa joined the Environmental Protection Agency (EPA) as a student and helped found the EPA's Office of Environmental Justice. He worked there for twenty-four years and says that there are "huge sets of challenges in front of us from the climate crisis" and "serious concerns with global warming, but I also see innovation and ingenuity. We can transform our society with new technologies to build a clean and just energy future. Things can be different. We can invest our time, bodies, and intellect—not to hope for change but to demand it."

In Mustafa's view, the most promising climate solution is environmental justice. "We can't win climate change if we don't address environmental racism," he said. "BIPOC communities have been hit hard by pollution. That's where fossil fuel refineries and toxic chemical plants are located; freeways, train tracks, and other transportation policies impact BIPOC towns too," he added. "In agriculture, decades of discrimination not only destroyed Black farming communities but meant we lost important lessons of regenerative agriculture." Mustafa underscored that "Black and brown folks are hurt the most from these environmental policies and practices. It's systemic."

His advice for those who want to be part of the climate movement is to "first, get educated. Use trusted resources, research, and learn. Second, get engaged. There are so many frontline organizations that need support." He also encourages self-reflection. "If you understand your unique gifts, you can better support those organizations that align with your interests." Finally, "Know where your dollars go. Where does your electricity come from? Where do your clothes come from? Let's better understand who we're supporting. Are we paying people to shorten our lives? How do we better use our individual resources to support the future we want to experience?"

Robyn O'Brien

Social Entrepreneur and Cofounder of Sirona Ventures

"I was such an unlikely crusader for cleaning up the food supply," said Robyn O'Brien, known as "The Erin Brockovich of food." A mother of four and self-described "Type A, firstborn child," Robyn graduated, on full scholarship, as the top female student in her MBA class. She took a job as a financial analyst for the food industry.

Robyn's surprising journey began when her youngest child was diagnosed with an egg allergy. This prompted her to dig into food allergies, including chemicals in food, the lack of transparency and regulation, and how food is grown. She wondered, *Are we allergic to food or to what's been done to it?*

In 2009 her findings became the bestselling book *The Unhealthy Truth: One Mother's Shocking Investigation into the Dangers of America's Food Supply—and What Every Family Can Do to Protect Itself.* Her book sent shockwaves through the food industry. "I learned that when you get that type of pushback, you have hit a talking point that [the industry] cannot address," Robyn remarked. It was "incredibly isolating and hard," not merely on Robyn but also on her family. Despite the stress, she said, "I would watch my kids play at the park and the other kids on the playground, and I knew that I had to continue."

Despite the food industry's pushback—or maybe because of it—word got out about Robyn's work. Her 2011 TED Talk has been viewed more than 1.2 million times. As the US public became aware that multinational companies were selling processed food with less harmful ingredients in other countries, the market shifted.

In true Beacon fashion, Robyn continues to be out front on cutting-edge issues. "The early part of my work was advising multinational companies that consumer demand for organic, real food wasn't a fad. It wasn't just moms being loud. What they were seeing was a fundamental change in how the twenty-first-century consumer was shopping, and that would disrupt their business model."

Eventually climate change—specifically soil health—emerged as Robyn's central focus. Another company she co-founded, rePlant Capital, finances farmers' transitions to regenerative and organic agriculture, which provides soil conservation carbon offsets for food company supply chains. "Conventional agriculture has practically killed the soil." Through regenerative agriculture, "we can return soil into the living sponge it's supposed to be that soaks up carbon and holds water."

Her message to companies is this: "As you heal the soil, you're taking care of consumer health, farmer health, and climate health." Robyn's message to people who want to get involved is to find your strengths, identify your passions, and get started, because even small changes can make a difference.

Gregg Renfrew

Founder and CEO, Beautycounter

"Climate anxiety is real. Of course, young people are worried, scared, and angry. As a parent, I tell my kids that the answers are out there, and I also acknowledge that these emotions are appropriate," remarked Gregg Renfrew, founder and CEO of the beauty brand Beautycounter. "But we have to take action. We can't be complacent. We have to use our cumulative brain power and move forward together."

Her Beacon Service Superpower reveals itself through entrepreneurship and dedication to transforming the beauty industry. Beautycounter has been recognized as one of Fast Company's most innovative companies year after year. Gregg's leadership disrupted the beauty industry and made reforming the 1938 federal cosmetics law a household issue. In December 2022, ten years after she and her team began lobbying for cosmetics reform, the Modernization of Cosmetics Regulation Act of 2022 was signed into law. As a Beacon Eco Hero, Gregg's creativity and confidence inspires consumers to live the change they want to see in the world.

"In 2006, I watched *An Inconvenient Truth* by Al Gore. It was a wake-up call. I was aware of global warming but realized I wasn't doing nearly enough and made the connection that what was detrimental to our earth was likely detrimental to our personal health. Gore's documentary helped inspire me to create Beautycounter. I knew from the start that my company would prioritize environmental health and sustainability."

With twenty years of experience working for prominent brands, Gregg launched Beautycounter and established the "never list" of toxic chemicals that the company wouldn't use in its product formulations. Gregg then recruited women across the country to join her as consultants to tell the story of safer ingredients, to sell safer products in their communities, and to lobby Congress to fix our broken toxics laws. The Carlyle Group announced a majority investment in Beautycounter and valued it at $1 billion.

Gregg sees extraordinary possibilities for businesses in climate solutions and in creating a regenerative economy. "Companies will have to recognize the challenges of the climate crisis, and many have. We can work toward a triple bottom line of people, profit, planet, because that's what consumers expect. Consumers want to be part of a community like that."

As Gregg explained, "Gen Z knows the challenges they're facing with the climate crisis. They are bright, and they understand that the solutions are right in front of us. Earth can heal itself if we let it. Ask yourself: How do you show up and how do you want to spend your time on earth?" One thing Gregg has learned is that "the small things matter. We have to take control of what we can control."

Her advice is to "pick a lane. Mine was transforming the beauty industry by eradicating toxic chemicals in cosmetics. I decided to concentrate on the environmental health movement." Even as an Eco Hero, Gregg recognized that "you can't do everything. Find your superpower to make an impact. Focus on that."

WHAT NEXT?

If you identify as a Beacon, do a gut check. Does this description resonate with you? If yes, let's put your Service Superpower to work. You can start right now. Grab a pen or pencil.

Three talents I have as a Beacon are:

1. ______________________________
2. ______________________________
3. ______________________________

Three things the climate movement could do a better job of are:

1. ______________________________
2. ______________________________
3. ______________________________

After reflecting on my interests and talents, here are three ways I might be able to help:

1. ______________________________
2. ______________________________
3. ______________________________

The 72-Hour Check Back

In 72 hours, pick up this chapter again. Write down one thing you learned about climate action over the past three days:

____

Now share what you've learned on social media or tell a friend or family member.

Then call your member of Congress and urge them to support strong climate action or a specific climate solution. The Congressional switchboard number is (202) 224-3121. Enter your zip code, and they'll connect you to your senator's or representative's office.

BEACON ONE GREEN THING EXAMPLES

Micro Impact

- Google "Cancer Alley" and learn more about environmental justice, environmental racism, and the impacts of the petrochemical industry on BIPOC communities.
- Join or sign up for the newsletter of a voting rights protection organization to ensure free and fair elections and full participation of citizens in our democracy.
- Have a conversation with a family member or friend who thinks there's nothing they can do to make a difference and persuade them using what you know about the climate crisis.

Macro Impact

- Find out when your local, state, or federal legislator is in town. Request a Zoom or in-person meeting with their staff to let them know that you demand action on clean energy.
- Go on your local television station, write an op-ed, or join a local podcast to share your opinions on the need for climate action.

If you're ready to get started, check out this 21-Day Kickstarter Plan for the Beacon to inspire your daily One Green Thing. Remember that the Laws of Simplicity & Consistency, Identity, and Amplification mean that your simple daily actions compound to ease eco-anxiety, shift the culture, and promote comprehensive solutions to the climate crisis. Note that the plan aligns with part 2, "The Seven Areas to Effect Change."

Feel free to either review the other profiles or skip ahead to chapter 10 for a visualization exercise and chapter 11 to check out the tools, including the Eco-Action Plan, Eco-Impact Top Ten, and Joy Tracker. Then move to part 2 to put your Service Superpower into action.

Be kind to yourself and remember that you're a product of a system that shifted disposal and energy emissions to you, the consumer, instead of the manufacturers who made all this stuff. The daily ritual of a One Green Thing is about intention and cultural change, not blame or shame. Climate activism isn't about perfection—it's about intention, personal growth, and daily action.

Be kind to yourself and remember that you're a product of a system that shifted disposal and energy emissions to you, the consumer, instead of the manufacturers who made all this stuff. The daily ritual of a One Green Thing is about intention and cultural change, not blame or shame. Climate activism isn't about perfection—it's about intention, personal growth, and daily action.

THE BEACON

21-DAY KICKSTARTER PLAN

	SUNDAY *know you can't go it alone*	**MONDAY** *think beyond your age*	**TUESDAY** *see energy in a new light*
WEEK 1	Start an honest conversation with a friend or family member about climate action and your concerns for the future.	Ask a member of another generation for a book recommendation. Discuss how that piece of literature impacted them and inspired change.	Research climate justice and donate to clean energy programs in lower income communities.
WEEK 2	Create a presentation on an environmental issue you're passionate about. Share at your place of worship or community organization, or with friends on social media.	Watch an environmental documentary you enjoyed with your family or friends. Invite someone younger or older than you to join. Lead a discussion afterward.	Challenge your office, friends, or community organization to see who can avoid using cars for one week through walking, biking, or public transportation.
WEEK 3	Get a group of friends to go to your next school board meeting. Testify about the importance of climate change education and readiness in school curriculum.	Take ten minutes to learn more about Fridays for the Future. Plan to attend a Gen Z climate rally or climate justice event.	Give a talk at a local school or community organization, or post online about climate solutions.

WEDNESDAY *understand that you are what you eat*	THURSDAY *protect the source*	FRIDAY *vote with your wallet*	SATURDAY *love your mother (earth)*
Google "regenerative agriculture" and learn how it can help create a better future. Write an op-ed for your local paper about it.	Challenge your friends to carry a reusable water bottle.	Start a conversation about a sustainable brand you support.	Research the history of your favorite national park or natural area. Discover its connection to Native American culture and share what you learn.
Write an op-ed or email your member of Congress about promoting sustainable agriculture and conservation.	Look up your water utility's next board meeting. Make a plan to go with friends and express your support for strong drinking water protections.	Host a backyard clothing swap and share the environmental benefits of using secondhand clothing.	Take a short walk outside. Engage all five senses as you connect with nature.
Research food justice and food deserts. Support a local organization that is creating better access to healthy food for BIPOC communities.	Encourage friends to join you in donating time or money to a water conservation program or favorite nonprofit.	Call the consumer hotline for a company that you think needs to switch to more sustainable packaging.	Advocate for accessible green space in your community by contacting local leaders and educating friends.

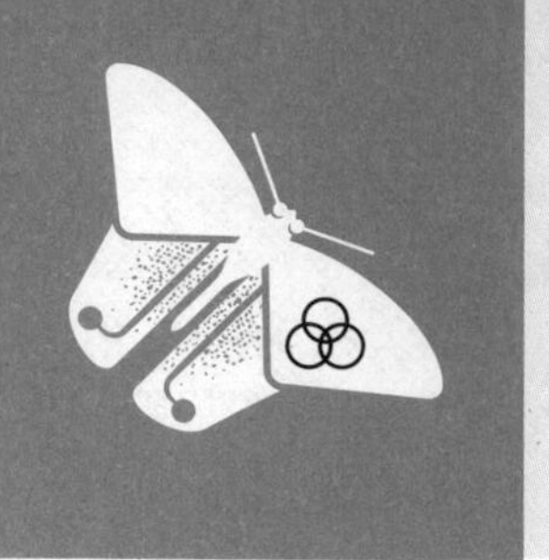

Chapter 5

THE INFLUENCER

THE INFLUENCER PHILOSOPHY:
"Connect. Share. Network."

INFLUENCER ATTRIBUTES AND STRENGTHS

Attributes

Influencers are sharers at heart. It's not enough to be a trend-setter. People with this Service Superpower want to connect people to their cause and inspire them to act, grow, and change their behavior. Influencers start their own initiatives, encourage others to attend events, share updates on breaking news and research, and introduce friends to people and ideas. They are charismatic, joyful, and powerful change agents.

Strengths

Strengths of the Influencer include an enthusiasm for learning; a desire to exchange ideas and motivate others to join a cause; a deep respect for people from all backgrounds and with differing views; a belief that sharing information, resources, and networking will inspire people to develop their potential; and a willingness to listen to others and swing into action if asked to help.

Influencers are visual and aural learners. They remember snippets of conversations, speeches they've heard, headlines they've read, and documentaries they've seen. They can quickly map out the relationships and possibilities between various subjects and people.

Because of this particular learning style, Influencers have a unique ability to remember people: their names, faces, and details about their lives. They thrive on connecting people to causes and to events and to each other. Influencers are fiercely intelligent, especially in motivating others, encouraging them to serve, and understanding how people interact. They're outstanding communicators and appeal to a wide range of audiences. They have a magnetism and energy that radiate innovation, compassion, and openness, and they possess an unmistakable "cool" factor. For these reasons, as well as their persuasion skills, outward focus, and others-oriented service, Influencers are people magnets.

Perhaps because Influencers are so focused on what makes others tick, they have a unique ability to predict and see trends in a wide range of topics, including but not limited to culture, sustainability, education, and science. Their extensive personal and professional networks mean they have their finger on the pulse of what will resonate with people and can help others identify causes that align with their values. This skill set matches service leadership well because the Influencer can create welcoming and inviting spaces for others to step up in service in the climate movement.

INFLUENCER CHALLENGES

The key areas for improvement for the Influencer are to increase their inward focus, advocate for themselves, and be willing to go along for the ride.

Increase their inward focus. Influencers tend to be outwardly focused. The excitement of connecting with others and serving the community with their full heart and soul means that they sometimes need to lean into the quiet of the Wonk and the Sage. Influencers must check in with themselves and ask what they need. *Have you eaten nutritious food today? Had enough water to drink? Moved your body? Meditated?*

Advocate for themselves. The Influencer is sharing, introducing, networking, reading, and listening, which means that sometimes what they want comes second. As they move from person to person or activity to activity, it's easy for personal needs to be pushed to a back burner. Influencers are well-liked but tend to be "pleasers" and can become resentful if they don't advocate for themselves. This book is all about service, but Influencers can quickly lose sight of their own objectives. They need to step back periodically and reflect and align on their individual purpose, mission, and goals.

Be willing to go along for the ride. The Influencer also needs to channel their inner Spark every now and then and simply enjoy the experience. In addition to feeling like they have to "be in the know," Influencers tend to assume they must organize everything for everyone. "Laid back" doesn't describe the natural state of the Influencer, but this profile can benefit from letting others take the lead and moving into a cheerleader role. Taking a pause can help restore their enthusiasm and fuel them for further action.

INFLUENCER OVERVIEW

Traits: *charm, eloquence, insight, open-mindedness, versatility*
Strengths: *magnetic, trendsetting, innovative, persuasive, popular*
Challenges: *overshares, pleaser, controlling*

STRATEGIES FOR INFLUENCERS TO RESTORE AND REPLENISH

Influencers are especially prone to burnout because so many people turn to them for inspiration, guidance, and next steps. Influencers need to restore and replenish through solitude, leaning into their Wonk, and delegating activities now and then. Here are three suggestions for Influencers to consider as they tend to their own eco-anxiety and exhaustion.

Ask for help. Delegate a task or ask a friend to come over and help with something—setting up a garden, planning a meal, or repairing something at home. A One Green Thing principle is "know you can't go it alone," which is sometimes hard for Influencers because so many people look to them for guidance. They can benefit from picking a task that they haven't been able to tackle and reaching out for help with it. If it happens to be a One Green Thing, that's even better.

Take a mental health day and spend it outside. Because Influencers are constantly connecting with others, they're usually talking, listening, caring, and attending. They should consider the revolutionary act of clearing their schedule for a day. Going outside is a powerful way to gain perspective and replenish energy reserves.

Focus on sleep. This might seem like a strange suggestion for a book that's trying to engage people in climate action. How can anyone sleep when the future of the world is at stake? The answer is simple: because we must. The Influencer would benefit from focusing on sleep hygiene for at least a week. Set a time for bed each night. Put away electronics at least one hour before bed. Meditate, pray, or journal to allow for a clear mind and better-quality rest.

Influencer Eco-Hero Profiles

Andrea Ambriz

General Manager of Exposition Park

Former Deputy Director, California Natural Resources Agency

"I think people need to know that you can change systems and policies by showing up, doing excellent work, and getting involved in meaningful ways," explained Andrea Ambriz. Named one of the "40 Under 40: Latinos in American Politics" by *Huffington Post,* Andrea now serves as the General Manager of Exposition Park, one of the largest open spaces in the heart of Los Angeles, and a key destination as the city prepares to host the 2028 Olympic Games. Andrea is the former head of External Affairs for the California Natural Resources Agency (CNRA). With more than a $7 billion annual budget, CNRA manages public lands, forests, state parks, water and coastal resources, wildlife, and wildfire protection for the state of California.

As an Influencer-Wonk, Andrea believes that convening individuals from different backgrounds unlocks multiple intelligences and creates paths forward. "That interdisciplinary approach of bringing people together—youth groups, faith-based organizations, climate justice leaders, and corporations—can't be 'a check the box' exercise. It's essential work to me." Andrea has strong policy chops, which means that as much as she gains energy from interacting with people, she loves to dig into the data.

She approaches natural resource issues through an intersectional lens of climate and economic justice. A former Obama White House official and presidential appointee at Treasury, Andrea returned to California to serve the communities she first supported as a young staffer in the state legislature. "My initial work in economic and community development brought me into environmental justice and conservation." She drafted an innovative grant program to invest in underresourced communities, called "The Statewide Park Development and Community Revitalization Act," which was signed into law in 2007. More than fifteen years later, Andrea's work still has ripple effects. When she visits pocket parks, playgrounds, and open green spaces across the state, Andrea says that's exactly what she envisioned. It's been a full-circle moment for her to see these investments pay off, as a leader of the team that oversees the state's natural resources.

Andrea believes that her role as an Influencer is to recruit more people into the climate movement through early education organizations, cultural institutions, and other innovators. She encourages everyone to learn the facts about climate change, then talk about it. "Make sure you call others to action. Reach out to people you know in your community or in government. Sign petitions. Show up to board meetings. Write letters," she urged. "Even if it seems daunting at first, get involved."

To create a bold, sustainable future we must "understand that environmental justice is linked to economic justice. We must invest resources in communities hurt by environmental injustice to recover, repair, and protect them from climate change. Future generations also deserve that opportunity as a down payment toward building a new generation of community opportunity and wealth."

Julia Cohen

Cofounder, Plastic Pollution Coalition

"What solidified my understanding of the connection between health and the environment was my mom's passing from breast cancer when I was just thirteen years old and she was only forty-three."

Julia Cohen is a cofounder and the managing director of Plastic Pollution Coalition (PPC), a global alliance of organizations, businesses, and thought leaders working toward a more just, equitable world free of plastic pollution and its toxic impact. She collaborated with her sister, artist Dianna Cohen, to create PPC in 2009.

"I had the policy and public health background; Dianna had the art and creative experience," Julia said. After a meeting at Google, Dianna discovered that there were many others focused on the global onslaught of plastic pollution.

"We wanted to create a big tent coalition where groups around the world could collaborate and gather to talk about and take action to stop plastic pollution." Julia had extensive experience in organizing at the State Department, Youth Vote Coalition, and Rock the Vote; it was natural for her to join Dianna in founding PPC.

Julia works with movers and shakers at all levels of government, sustainable business, and the entertainment industry. She's a powerful networker who constantly introduces people to each other and follows and shares the latest trends in science, environmental health, and climate action. Julia realized her Service Superpower later in life. "But it makes sense to me," she said. "My mom ran the LA Free Clinic, then the Hollywood Human Services Project, and my dad is a documentary filmmaker and author. My family focused on art, service, science, and connecting."

Her advice to entrepreneurial Gen Zers and Millennials interested in climate action is to "check first to see what's already going on and who's out there in your community taking action. Drill down and do the research, and get involved."

For those new to the climate movement, she advised, "Identify what you care about. Get educated, but don't get overwhelmed, because there are millions of rabbit holes to get lost in."

Julia also suggested reading science fiction "because we have to dream it before it happens." She highly recommended the bestseller *The Ministry for the Future* by Kim Stanley Robinson, which envisions a future in our warming world.

Julia pointed out that "deep down you have to be an eternal optimist to run a global coalition." She called the millions of people engaged in and fighting for climate action over the past few years "amazing." She's also inspired by how the Black Lives Matter movement, Fridays for Future, and Climate Action are coordinating and actively working to address climate justice.

The Break Free from Plastic Movement, of which PPC is a founding member, has taken hold in Europe, China, Southeast Asia, and Africa as well. "It's a testament to the power of thinking big and speaking truth to power," said Julia, "and Plastic Pollution Coalition is proud to be a part of it."

Robin Hill-Emmons

Social Entrepreneur

Robin Hill-Emmons exemplifies the Influencer Service Superpower profile, with a strong Philanthropist wing. Robin started the nonprofit food-justice organization Sow Much Good after her brother, who struggled with mental health issues and homelessness, needed help. After becoming his legal guardian and finding him proper care, Robin was surprised that her brother became borderline diabetic, even while his mental health improved. Robin realized that the processed food the housing facility provided him was the culprit. To her shock, he'd received better nutrition when he was homeless. Robin left her corporate job, tore up her manicured suburban yard, and planted a garden of fresh vegetables for his facility.

"I used social media to tell people about my garden. More than fifty people showed up and helped me harvest. I told them that I wanted to feed people in a transitional housing group, and one of those people was my brother," Robin explained. "When neighbors would ask why there were so many cars parked on our street, I asked them to join us. And they did."

Robin says that a major component of being an Influencer is believing in the power of compassion and believing that people want to create a better world. Sow Much Good expanded quickly, from supporting Robin's brother's facility to targeting the BIPOC community in low-wealth areas, creating a community-supported agriculture program, developing a farmers market, offering cooking classes, and sustainably growing more than twenty-six thousand pounds of food on more than nine acres in and around Charlotte, North Carolina. Robin's advocacy garnered significant media attention, including features in *People*, CNN, PBS, *Modern Farmer*, and *Southern Living*. Robin also received the CNN Hero Award for her leadership.

One of her key lessons in advocacy is that "the heart can't feel what the eyes can't see." She understands that homelessness, food justice, and climate change are interrelated on many levels, including the need for people to "see" the challenges.

In 2018, after eleven years of service, Robin wound down the nonprofit. By then other food justice groups had emerged to continue the work. "Through this experience, I knew that I could be a leader of others to create big solutions." Robin always fights for the underdog with her "finely calibrated moral compass." Systemic racism and the challenges of growing up in a low-wealth community motivated her to be a bright light for positive change.

Since founding Sow Much Good, Robin has traveled the world, including to China on the prestigious global Eisenhower Fellowship for agriculture and the environment, and now advises nonprofits and philanthropies on sustainability and racial equity. Robin's advice is to know your strengths and "start where you are," quoting Dr. Martin

Luther King Jr.: "'You don't have to see the staircase—just take the first step.'" She then added, "Plant a garden or a tree. Participate in a watershed cleanup. Just begin and invite others to come."

WHAT NEXT?

Once you've read the strengths, attributes, areas for development, and ways to avoid burnout, ask yourself if the Influencer Service Superpower profile reflects who you are. In true Influencer fashion, ask a friend or family member if these traits and talents sound like you. Then make a list of how you think you can use your traits to help the climate movement:

My top three strengths as an Influencer are:

1. ________________________________
2. ________________________________
3. ________________________________

Three things the climate movement could do a better job of are:

1. ________________________________
2. ________________________________
3. ________________________________

After reflecting on my interests and talents, here are three ways I might be able to help:

1. ________________________________
2. ________________________________
3. ________________________________

The 72-Hour Check Back

In 72 hours, come back to this section and reflect on the Influencer profile and your One Green Thing journey. What's working? What are you excited about? What do you want to try next? What environmental or climate issues do you want to learn more about?

____

____

____

Now call or text a friend or family member and tell them that they, too, can join the climate movement. Share that you're embarking on a daily practice of sustainability and that you want them to do it with you. Enlist three people. Write their names here:

1. ____
2. ____
3. ____

INFLUENCER ONE GREEN THING EXAMPLES

Micro Impact

- Buy secondhand clothes instead of fast fashion.
- Learn and post about climate justice.
- Make a plan to attend an environmental event, like a music festival, art show, or the National Day of Unplugging (complete with a cell phone sleeping bag!) or opt for a real sleeping bag and participate in the National Wildlife Federation's Great American Backyard Campout.
- Encourage friends to attend Earth Day celebrations and lobby days with you.

Macro Impact

- Pick your favorite climate solution or One Green Thing (see the Kickstarter Plan and part 2 for ideas) and invite three friends to do it with you.
- Learn how to lobby members of Congress about the need for environmental education, and then recruit others to do the same.

If you're ready to get started, check out this 21-Day Kickstarter Plan for the Influencer to inspire your daily One Green Thing. Remember that the Laws of Simplicity & Consistency, Identity, and Amplification mean your daily actions compound to shift the culture and promote comprehensive solutions to the climate crisis. Note that the plan aligns with part 2, "The Seven Areas to Effect Change."

Feel free to either review the other profiles or skip ahead to chapter 10 for a visualization exercise and chapter 11 to check out the tools, including the Eco-Action Plan, Eco-Impact Top Ten, and Joy Tracker. Then flip to part 2 and put your Service Superpower into action.

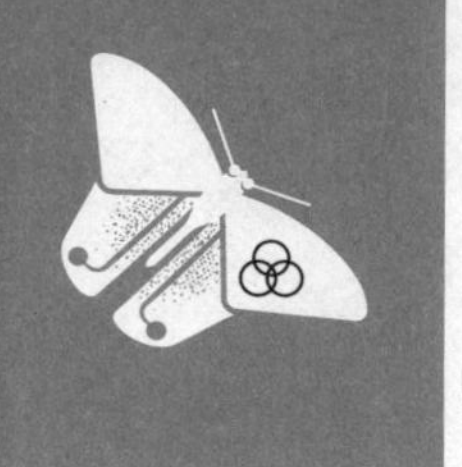

THE INFLUENCER
21-DAY KICKSTARTER PLAN

	SUNDAY *know you can't go it alone*	MONDAY *think beyond your age*	TUESDAY *see energy in a new light*
WEEK 1	Advocate for accessible green space in your community by contacting local leaders and educating friends.	Host an intergenerational book club that discusses current events, climate change, technology, and the future.	Share information about your favorite climate solution (e.g., weatherizing your home, community solar gardens, etc.).
WEEK 2	Encourage friends and family to talk about what compassion means to them.	Tell someone in another generation about your feelings on the climate crisis. Ask them to share what they know.	Write down three ways you could save energy, and educate your family or friends on how they can help.
WEEK 3	Look for examples of compassion in your community, take a picture, and share on social media (e.g., a community garden).	Talk with someone who remembers the Nuclear Freeze Movement, or share your experience with a younger person. Discuss how positive change is possible.	Call or email your utility and ask about its clean energy portfolio. Share what you learn on social media. Be sure to use the hashtag #onegreenthing.

WEDNESDAY *understand that you are what you eat*	**THURSDAY** *protect the source*	**FRIDAY** *vote with your wallet*	**SATURDAY** *love your mother (earth)*
Try out a new plant-based recipe and post it on Instagram or TikTok, or encourage friends to start a plant-based potluck or recipe exchange.	Use your social media platform to share information on organizations that support clean water and reduced water usage.	Make a list of the top five sustainable products you buy and encourage your friends to check them out.	Take ten minutes to research an endangered species. Share what you learn on social media. Email the appropriate agency to urge them to protect its habitat.
Share about the importance of soil health and its relationship to the foods we eat.	Call the mayor's office or your city council to urge them to protect local streams and waterways.	Give sustainable gifts promoting green living, or skip gifts and choose experiences or donations over stuff.	Share resources on how to support public lands and how development pressures and climate change are impacting wildlife.
Inspire others to think about food waste in new ways—composting or menu planning to use leftovers and reduce waste.	Look up the water quality in your zip code. Urge your mayor, city manager, or other local leader to provide strong funding to ensure clean drinking water.	Declutter and then "audit" your home to see if you can make more sustainable swaps for household items.	Google the best walk, park, or birdwatching spot in your neighborhood. Plan a trip and encourage your family and friends to join you.

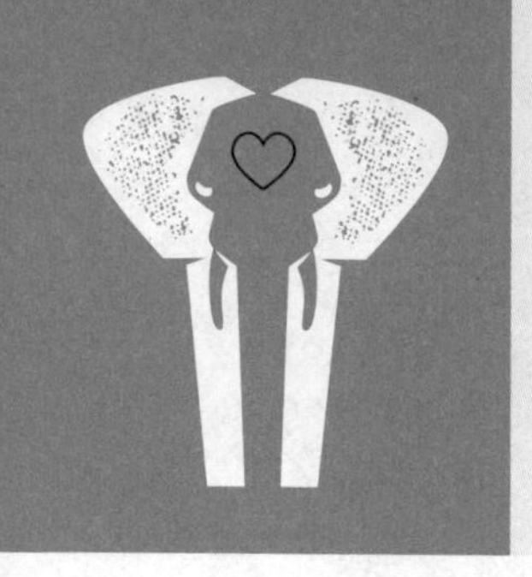

Chapter 6

THE PHILANTHROPIST

THE PHILANTHROPIST PHILOSOPHY:
"Give back. Reach out. Provide opportunity."

PHILANTHROPIST ATTRIBUTES AND STRENGTHS

Attributes

The Philanthropist possesses an extraordinary drive to empower others, zeal in bringing people together, an ethic of service to inspire friends to give their time and resources, and a profound sense of the wonder of nature. The Philanthropist serves a critical role in accelerating climate action by providing time, talent, and financial resources. This profile sees possibilities, introduces others, and loves to sign up for causes. As a giver, this Service Superpower profile gets energy from being with people, learning about game-changing ideas, and seeing others' dreams come to fruition. For some people the word *philanthropist* may conjure up images of the Monopoly game guy with the monocle, mustache, and top hat, but the word means "one who makes an active effort to promote human welfare."[1] For this profile the active effort includes volunteering, connecting, and donating. Anyone can be a Philanthropist because this service profile is about contribution.

Strengths

Philanthropists are confident, engaging, and curious. They are big-picture thinkers, innovative, and inspirational in that they know how to make things happen. They're also strategic and thoughtful about where they invest their resources. Since they are relationship-focused, they have a great memory when it comes to people, their stories, and their causes. Philanthropists ask probing questions because they want to get to know people, create deep relationships, and make strong connections over time.

Traditionally highly sensitive and empathetic people, Philanthropists are change agents. Their networks and resources can transform ideas into realities. They "read people" well and know how to motivate groups. They define "return on investment" in terms of shaping social impact, moving exciting concepts and technologies forward, and emboldening other leaders and communities to reach their potential.

Filled with gratitude, Philanthropists seek ways to empower others and gain satisfaction in others' success. Much of their work is about amplifying research, breaking down silos, and weaving networks of people and ideas to create a better world. Given their positive outlook on life, Philanthropists tend to be "happy warriors" for social change.

Even though the Philanthropist Service Superpower profile expands beyond financial contributions, most Philanthropists are comfortable around money. Whether making money, asking for it, or investing it in social causes, philanthropists don't have financial hang-ups. With their abundance mentality they see supporting others and causes that are important to them as a way of paying it forward, meaning that providing opportunities to others sends more positive energy into the world.

PHILANTHROPIST CHALLENGES

This profile can focus on three areas of development: avoid burnout, manage commitments and time, and go with the flow.

Avoid burnout. Because of the myriad demands on their time, Philanthropists are prone to burnout. To avoid this, they need to create space for themselves to reflect and prioritize. As givers, they want to support everyone, but that's impossible. Scheduling non-negotiable quiet and wellness time, such as exercise, massages, and other self-care practices, can prevent burnout.

Manage commitments and time. The Philanthropist tends to over-commit because they're enthusiastic about change and buoying different causes. Their generosity extends to their calendars, but they need to streamline their service activities and manage their obligations closely. Sometimes Philanthropists can seem distracted because they have so much going on in their brains, and many people are vying for their time and attention.

Go with the flow. Because this Service Superpower profile is all about giving back, Philanthropists dedicate a lot of time to their communities. However, they tend to like things done a certain way. A strict view of how to make change or how to complete a project can sometimes be counterproductive. This Service Superpower profile needs to let go of their fixed idea, let others lead, and trust that it's going to be okay. Letting go can be especially challenging when you're dedicating time, expertise, or financial resources. That's why it's important to underscore that the Philanthropist's service isn't necessarily directing—it's giving back and restoring.

PHILANTHROPIST OVERVIEW

Traits: *graciousness, strong work ethic, kindness, pragmatism, acumen*

Strengths: *generous, connected, selfless, savvy, imaginative*

Challenges: *controlling, overcommitted, distracted*

STRATEGIES FOR PHILANTHROPISTS TO RESTORE AND REPLENISH

Philanthropists can restore by leaning into their Adventurer and Spark sides. Because this profile wants to fix everything and give to everyone, they sometimes need to drop their responsibilities for a few days and go on a big adventure or get outside their comfort zone. Here are three ways to replenish:

Go on a spontaneous adventure. For the Philanthropist, canceling all calls and meetings for the day is the type of jolt that can generate more positive energy. Seeing a friend in a nearby town, floating down a nearby river, or visiting a local nature center can shift the Philanthropist's perspective.

Journal in nature. Philanthropists can benefit from this simple exercise: Bring a pad and paper and go out into nature. Engage all five senses: listen, look, smell, feel, and taste.

Write down all feelings for ten minutes without stopping, a practice known as free writing. Don't worry about spelling, grammar, or punctuation—just write whatever comes to mind, without censoring or judging.

Write down three top priorities and check in monthly. A Philanthropist can gain clarity by spending thirty minutes in deep thinking, pondering questions such as the following: Why am I focused on service? What in my background inspires me to serve? What type of systems change can I help support? What do I need to work on myself?

After they describe the three areas, organizations, or issues they will support, they should check back monthly to ensure their goals align with their schedule.

Philanthropist Eco-Hero Profiles

Jennifer Caldwell

Foundation President and Nonprofit Leader

"From a young age, I have felt very connected with nature—and inspired by it. My mom would frequently discover me with a sketch pad, drawing an animal or plant," remarked Jennifer Caldwell, president of the Caldwell Fisher Family Foundation and board member of The Nature Conservancy in California and Project Drawdown. This connection to nature translated into a life of service in conservation, climate action, and the arts. As a Philanthropist Eco Hero, Jennifer tackles global problems by empowering others, asking the right strategic questions, and creating dynamic networks of change agents.

When Jennifer realized her "true passion was the environment," she took her private-sector publishing, media, and broadcasting career from New York and Silicon Valley to head marketing at The Nature Conservancy, which focuses on biodiversity and land conservation.

Al Gore's documentary *An Inconvenient Truth* "had a profound effect" on Jennifer. "I was an environmentalist. I had worked for one of the top environmental organizations in the world and somehow, I was just learning about this extraordinary threat. It felt imperative that I do something," she explained.

Jennifer then created a movement called "Hope to Action" that focused on women "whose moral authority and consumer power, I believe, are potent drivers for change." While she was bringing up two toddlers, "climate change had risen up like a phantom, casting an even greater, more pronounced and worrisome concern to our global environmental health." Several years into this endeavor, Jennifer shifted gears and merged her organization with another nonprofit.

To decide where to dedicate her time and resources next, Jennifer "started to map out all the sectors for possible climate action, from energy to population control, to nature-based solutions." When she discovered the book *Drawdown,* she says, "I knew I'd found my road map." The book inspired her because "it declared that it was time to move past hand-wringing over the dire threat of climate change, toward solutions. That deeply appealed to me as a hopeful person inclined toward action."

Her advice on where to begin includes: "Every job is a climate job: consider what contribution you can make from your platform. Everyone has something to contribute." Jennifer also urged readers to "empower and educate those around you—whether they be family or friends—on how they, too, can shift to 'the right side of history.' We all have a stake in this."

What keeps Jennifer focused on the positive is that "climate change is acknowledged by the majority in our country, and there's no reason why our future when it comes to our planet and its health should be politicized," she explained. "What gives me hope is action. I try every day to overcome inertia. Citizens will make the right choice if they know what the right choice is and if made economically viable and easy for everyone."

Maya Penn

Fashion Designer, Animator, and Sustainability Consultant

Profiled by *Teen Vogue*, CNN, and *The View,* twenty-four-year-old entrepreneur, philanthropist, three-time TED speaker, sustainability consultant, artist, global activist, coder, and author Maya Penn has inspired people all over the world. She started her sustainable fashion company, Maya's Ideas, when she was eight years old. Her creative genius, giving nature, and keen intelligence make her a Philanthropist Eco Hero.

"The more I learned about sustainable fashion and the environment, the more ideas I had. I knew I wanted to work to support environmental justice and humanitarian efforts, so I founded my nonprofit, Maya's Ideas for the Planet, in 2011, when I was twelve years old. Projects include providing eco-friendly sanitary pads to girls in Haiti, Senegal, and Cameroon. In her hometown of Atlanta, Maya's nonprofit distributed masks and food during the pandemic. "It's been amazing to see the impact in the US and around the world."

As a Philanthropist, Maya says that giving back is part of how she was raised. "My earliest memories are dropping off canned food and clothing at shelters. My parents always believed that we all need to play our part in some way to create a better world."

As a self-described "really nerdy kid," Maya has always been exceptionally artistic and had an affinity for animals and nature, which inspired her research on the fashion industry's environmental impact. At first "a lot of my work included education because sustainable fashion wasn't mainstream back then. People didn't understand the impact that the industry has on water, toxic chemicals, greenhouse gas emissions, and labor." This led to Maya's first TED Talk at age twelve, more speaking engagements, and her book, *You Got This: Unleash Your Awesomeness, Find Your Path, and Change Your World.*

Maya's advice to young people is "to focus on positive action, not the negative impacts that surround us. We didn't ask to be part of this destructive system, but we're part of the solution. You don't have to change the world. But you can think about change in your community, your home, your friend group—whatever that action looks like to you." She encourages people to find their why and to connect their talents to environmental action: "Of course, protecting the environment is the right thing to do, but what work connects with you? If you love the oceans, maybe that could be your focus—marine biology, blue carbon, or plastic pollution. I am an artist, so art is part of what I do. People resonate with my creative element—animation, film, and design always come into my work."

Maya wants young people to know that their voice makes a difference. "Find something you care about. Beyond the need for awareness, we have to focus on hope and solutions," she said.

Laura Turner Seydel

Chair Emeritus, Captain Planet Foundation

"I'm the ultimate volunteer," said Laura Turner Seydel, who sits on more than a dozen nonprofit boards ranging from climate action to international peace efforts. As a Philanthropist Eco Hero, Laura loves to promote game-changing environmental solutions. She's a sought-after speaker on sustainability, children's health, and empowering people to lead change for a sustainable world.

Regenerative agriculture brings Laura hope. Ted Turner, Laura's father and the creator of CNN, owns just under two million acres of land in the United States. As a ranching business, the family decided to invest in soil health to foster grasslands restoration and bison recovery.

"The more I've learned about the opportunities for soil regeneration to address the climate crisis, the more optimistic I've become," explained Laura. "I've witnessed firsthand how implementing regenerative agriculture and adaptive grazing creates win-wins for people, animals, and the planet. We need to rethink the harmful farming methods that have reduced biodiversity, increased drought conditions, and polluted the environment. Transitioning to regenerative practices can build topsoil; increase profits, biodiversity, and drought tolerance; and draw down greenhouse gasses."

Laura mentioned research showing that healthy grasslands sequester twenty times more carbon than trees in forests and that "nature is resilient when you pay attention to her." She shared that *The Biggest Little Farm* and *Kiss the Ground* are two essential documentaries that show how regenerating soil is a top environmental solution.

Even though individual action, like recycling and reducing throwaway plastics, can make a difference, Laura said it's unfair that "certain companies have put the onus on consumers for the plastic pollution crisis. Companies should be accountable for the costs of the pollution caused from plastic packaging. According to the EPA, only 8 percent of plastics are recycled in this country. I have witnessed intentional tactics to keep people confused about how and what to recycle. Plastics are choking our rivers, killing wildlife, clogging sewer systems, polluting people's bodies—even infants. The climate crisis is similar. The fossil fuel industry has known about the science on climate change for decades. Their greed is leading to runaway climate change. But Project Drawdown enlisted researchers to identify the top scalable solutions to global warming." A few of her favorite solutions are reducing food waste and eating plant-rich diets.

Laura recommended volunteering as an antidote to stress and anxiety. "Whether it's a river cleanup, planting a community garden, or making meals for the homeless, service is a healing experience, knowing that you can make a difference."

Laura also highlighted the experience of her stepmother, Jane Fonda, who initially found her climate despair overwhelming. Then Fonda created the Fire Drill Fridays campaign with Greenpeace to protest for global climate policy. "In taking action and fighting for climate justice, she found her joy," remarked Laura.

WHAT NEXT?

Now that you've read the strengths, attributes, areas for development, ways to avoid burnout, and the Eco-Hero profiles, does the Philanthropist Service Superpower seem like the right fit for you? If so, list your strengths in service:

My top talents as a Philanthropist are:

1. ______________________________
2. ______________________________
3. ______________________________

Three things the climate movement could do a better job of are:

1. ______________________________
2. ______________________________
3. ______________________________

After reflecting on my interests and talents, here are three ways I might be able to help:

1. ______________________________
2. ______________________________
3. ______________________________

Now think of three areas you want to get involved in. What have you always cared about, even when you were a kid? Animal welfare? The arts and nature? Real food access? Spend ten minutes researching specific environmental issues or climate solutions that you're passionate about. (Feel free to check out part 2, "The Seven Areas to Effect Change," or the appendix for ideas.)

1. ____
2. ____
3. ____

The 72-Hour Check Back

Identify three organizations you want to learn more about and potentially support. Then commit to volunteering, donating, or sharing educational resources of one of these three organizations in the next 72 hours.

The three organizations are: Circle the action you took:

1. ____

Donated / Volunteered / Shared

2. ____

Donated / Volunteered / Shared

3. ____

Donated / Volunteered / Shared

PHILANTHROPIST ONE GREEN THING EXAMPLES

Micro Impact

- Feel inspired after sharing a conservation group's email newsletter.
- Volunteer at a phone bank with a local environmental organization.
- Create an online fundraiser for a local land trust.

Macro Impact

- Lead a multimillion-dollar campaign for an environmental education program at a university.
- Organize colleagues to urge your pension or IRA fund manager to divest from fossil fuels.

If you're ready to get started, read through the 21-Day Kickstarter Plan for the Philanthropist. Scan these ideas to inspire your plan. You can help shift the culture by applying the Laws of Simplicity & Consistency, Identity, and Amplification. Your simple, daily actions add up, can bring more joy to your life, and can promote comprehensive solutions to the climate crisis through culture change for a more sustainable future. The Kickstarter Plan mirrors part 2, "The Seven Areas to Effect Change."

Feel free to either review the other profiles or skip ahead to chapter 10 for a visualization exercise and chapter 11 to check out the tools, including the Eco-Action Plan, Eco-Impact Top Ten, and Joy Tracker. Then flip to part 2 and put your Service Superpower into action.

Stepping out of eco-anxiety into action takes more than hope. It's about courage. Acknowledge the feelings of stress and despair, lean into compassion, and commit to action every day. Stand up for the next generation and our shared future.

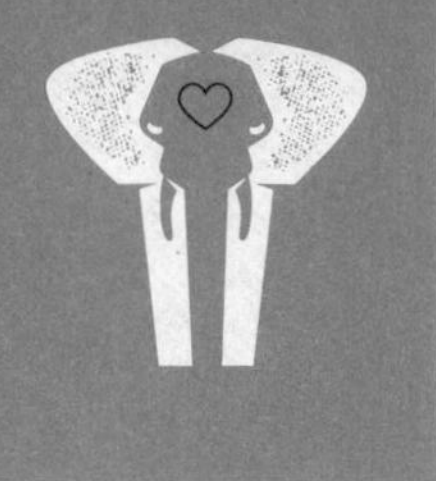

	SUNDAY *know you can't go it alone*	MONDAY *think beyond your age*	TUESDAY *see energy in a new light*
WEEK 1	Recruit three friends to call your member of Congress asking them to support strong federal climate policy.	Invite someone from a different generation to a fundraiser for an environmental organization you support or want to know more about.	Consider investing in an organization that provides clean energy to low-income communities.
WEEK 2	Raise funds for groups that promote access to the outdoors for BIPOC communities.	Volunteer with a youth-led organization or an elder-care organization. Learn what they think about the climate crisis.	Share information on social media about renewable energy technology and carbon offsets.
WEEK 3	Host an event to talk about compassion and the environment.	Connect a school environmental club with environmental professionals, or vice versa.	Share information about community solar and clean energy options in your area.

WEDNESDAY *understand that you are what you eat*	THURSDAY *protect the source*	FRIDAY *vote with your wallet*	SATURDAY *love your mother (earth)*
Make a plan to volunteer in a food bank, community garden, park, or local school program.	Give to an organization that champions clean water, or volunteer for a local river cleanup day.	Take fifteen minutes to learn about fast fashion. Write down three eco-friendly fashion options to explore.	Ask for friends and family to support public lands and national parks in lieu of gifts for your birthday or other holidays.
Research and support local food banks that accept fresh and perishable foods.	Spend fifteen minutes researching which reservoir your community water comes from. If you have a well, research well safety protocol.	If you have investments or a retirement account, spend twenty minutes checking if they are in fossil fuels. Consider supporting greener options.	Attend or host a fundraiser to support a local land trust or wildlife organization.
Support programs that connect local farmers to schools, restaurants, and food banks.	Share your commitment to reducing water usage on social media.	Share on social media why parks, public lands, and green spaces need support. Use the hashtag #onegreenthing.	Decide to give your favorite green products to friends and family for special occasions this year. If you don't have any top choices, ask friends and family about the green brands they love.

Chapter 7

THE SAGE

THE SAGE PHILOSOPHY:
"We are all connected."

SAGE ATTRIBUTES AND STRENGTHS

Attributes

The Sage has a deep spiritual connection to nature. This Service Superpower is driven by a compassion for others and a commitment to listening, engaging, and acting on environmental issues. Given their often-religious background, Sages seek consensus and common ground in communities. They're seen as healers by their peers and are clear communicators. Sages view conservation as a stewardship model, where they're caretakers of the planet in service of a higher power.

Sages need time alone for reflection, prayer, or meditation in nature. Time in the outdoors infuses them with peace and energy. Their concept of service involves people, but Sages are also motivated by protecting wildlife and animals in general. This Service Superpower profile has a strong intergenerational bond and yearning to create spaces for communication across ages, culture, and political views.

Strengths

Since Sages frequently transcend traditional categories that divide people on political fault lines, they have a unique ability to break through rancor and ideology and build bridges. Intuitive and wise, Sages have a knack for conflict resolution. With their compelling sense of purpose, they enable others to see common interests and engage with practical solutions. This profile forces people to step back, consider the larger picture, and appreciate community. Sages enable others to see the world outside of themselves and to witness the interconnectedness of our journeys here on earth.

In the context of climate action, Sages appreciate the awe of biodiversity, revel in the majesty of the natural world, and understand the effects of global warming. This Service Superpower profile feels a serious personal duty to protect nature. Their gifts in communicating spiritual lessons in service translate into an impressive ability to inspire behavior change. For example, Sages can tie compelling verses in spiritual texts to environmental stewardship and reframe environmental action from a political decision to an ethical duty. Evoking the universal principles of compassion, care for community, and assisting others means Sages encourage people to embrace long-term thinking (see chapter 13). Sages remind their communities of the intergenerational obligation to create a livable, thriving planet.

Sages are community focused. They understand the psychological need for belonging and the value of designing experiences and events for people to be heard, respected, and lifted up. To establish a just, equitable future, community-based solutions are necessary. As we construct a more inclusive climate movement and unlock everyone's best thinking, Sages assist in devising welcoming community spaces to shift conversations from hope to action. The Sage's belief that a higher power exists keeps compassion front and center as we adapt to the impacts of climate change.

The Sage's voice is crucial not only in consensus building but also in creating a spiritual case for action. This is particularly important with family members or friends who don't accept climate change. As climate scientist and evangelical Christian Katharine Hayhoe (a Sage-Wonk profile) reminds us, talking about the climate emergency "changes social norms." She stated: "And so the more we talk about it, the more we instill the idea that it is not acceptable to continue digging up and burning fossil fuels that are wrapping an extra blanket around our planet causing it to warm."[1]

Sages also restore others' energy as they take on overwhelming issues like global warming. Because their relationship to nature is a spiritual one, they tend to be more optimistic and hopeful than other profiles. Their rock-solid spiritual beliefs translate into a certainty that instills confidence and peace to others. Although the Sage's relationship to the environment is a personal one, sharing their wisdom and perspective provides comfort and emboldens colleagues to stay on course and not give up. The Sage's calm personality alleviates worry and stress for others.

SAGE CHALLENGES

Areas that the Sage can develop are to translate their personal relationship to nature into action, be flexible, and create room to express personal feelings of anxiety or overwhelm.

Translate the personal relationship to nature into action. Sages are attuned to their individual spiritual bond with nature. Time in nature and connecting to the outdoors are essential but must also be grounded in the impacts of global warming like extreme weather events, species extinction, and massive water and air pollution. To counter this inward-focused tendency, this profile may need to work on translating their personal relationship to nature into action. Sages can lean into the Wonk traits to share climate solutions. To ensure that their inner world is aligned with their outer activities, Sages should also lean into the Beacon every now and then. Speaking up and sharing their personal sense of duty to protect the environment will inspire friends and colleagues to follow.

Be flexible. The moral authority Sages command can comfort and sustain friends and family in difficult times, but their clarity can also make others uncomfortable. Despite their talent for bridge building, their certainty in their worldview can sometimes come across as fixed. Sages need to review their language and approach periodically to remain inclusive and inspiring in action.

Create room to express personal feelings of anxiety or overwhelm. Sages spend a large percentage of their free time in service. They are advisors, listeners, counselors, and wonderful friends. As a result, Sages must consciously design time and space for their own needs, including any anxiety and stress and their personal concerns about the future.

SAGE OVERVIEW

Traits: *humility, patience, active listening, wisdom, calmness*
Strengths: *spiritual, peaceful, thoughtful, fair, supportive*
Challenges: *inward, apolitical, inflexible*

STRATEGIES FOR SAGES TO RESTORE AND REPLENISH

Sages can burn out as they listen to and care for others who are worried about the future. More and more people are experiencing the mental health impacts of the climate crisis (see chapter 13 for more information). This mental stress, in addition to the physical stress from enduring extreme weather events, can exhaust this profile. Once engaged in the climate movement, the work can be tiring on a spiritual level. Here are ways for Sages to avoid burnout:

Draw, paint, or listen to music outside. Sages can connect with their spiritual side through the arts, such as sketching a familiar outdoor scene or a favorite pet or animal. Even fifteen minutes of these activities can help restore energy and increase a sense of well-being.

Get mad. This statement makes many Sages uncomfortable. They are even-keeled and known for having their emotions in check. Sometimes, though, it's good to let off a little steam in a safe, quiet, secluded place. Bearing witness to climate change impacts internationally and focusing on the disparate impacts on the under-resourced and Black, Indigenous, and People of Color communities can overwhelm even the most serene person. Doing a strenuous workout or ranting out loud for a few minutes can release the fear, anger, or frustration. Check out the Climate Mental Health Network's climate emotions wheel for guidance. The Sage can then head back to meditation, prayer, and calmness.

Write and share. Sages have a rich inner world. Sharing their wisdom and talent is an important service to other generations. Writing about their insights on the day's events can both ease their minds and inspire others. They can also take fifteen minutes and write their thoughts on climate action, spirituality, nature, or any of the themes in part 2. It may also be helpful to share their writing with someone from a different generation and ask for their views.

Sage Eco-Hero Profiles

Reverend Sally Bingham

Canon for the Environment in the Diocese of California

"I didn't see protecting the planet as a partisan issue; it was a religious issue for me. How we treat our neighbors comes from inside our soul; it's a spiritual issue," reflected Reverend Sally Bingham. Honored as a "Sacred Gift to the Planet" by the World Wildlife Fund and "one of the leaders of new green revolution" by *Rolling Stone*, her initial call to join the ministry came from her concept of "Creation Care," the spiritual duty to protect the earth. As a member of the Environmental Defense Fund board of directors, Sally met with the nation's experts on global warming.

"I was learning so much about God's creation and how we were not being good stewards. I kept asking my priest why the church wasn't talking about the environment. Finally my rector said, 'Why don't you just go to seminary?' So I did."

She felt called to go deeper into the possibility of being ordained, "but I didn't realize that you needed a college education to enroll in seminary. At forty-five years of age, I registered as a freshman at the University of San Francisco, got my degree, then went to seminary." She later founded The Regeneration Project to engage spiritual leaders and congregations in how to take care of God's creation. Her organization's central campaign became Interfaith Power & Light, with programming in all fifty states and more than 22,000 congregations involved. The work has been "incredibly rewarding" as she has engaged the Greek Orthodox community, Bahá'í,

Hindu, and Muslim faiths in climate action. "Yes, God gave us gas and oil, but God also gave us wind and sun. So why not use those God-given clean energy resources? Through Interfaith Power & Light, we've put solar panels on churches, mosques, synagogues," explained Reverend Bingham.

When she talks about climate, she always starts with an apology to young people "for what my generation is leaving: a really destroyed planet." Several of her congregants have decided not to have children because they don't think the planet will be healthy for the next generation. "I wish I had gotten involved sooner," she added.

After Reverend Bingham apologizes, she looks to all the action that is taking place now. "Renewable energy is now cheaper than coal. Fossil fuels are on their way out. Start with being conscious. Don't waste as much food. Eat less meat. Commit to make your next car an electric car."

She emphasizes that momentum is shifting, and more people want to get involved in climate action. When she first started talking about climate change as a spiritual issue, some congregants said that she was a "communist." Her response back then was simple: "This is not politics. It's life or death. If you say you're religious, then you have to be responsible to creation. God so loved the earth that he gave his only Son. It's our mandate from God to protect all that's been given to us."

Reverend Gerald L. Durley, PhD

Civil Rights Leader, Interfaith Power & Light Board Chair

"First of all, I'm not an Eco Hero. I am an 'eco-participant.' It's a real battle we're in. Never forget that we're on the winning team. It might be tough, and we'll encounter valleys and climb high mountains, but we will prevail," insisted Reverend Dr. Durley. Recently recognized on the National Park Service's "International Civil Rights Walk of Fame" for his leadership, Reverend Dr. Durley is a Sage Eco Hero and civil rights icon. He walked with Dr. Martin Luther King Jr. in the 1963 March on Washington. The author of *I Am Amazed!*, Reverend Durley refers to himself as a *rewired* pastor, not a retired one, given that his focus has shifted from a specific congregation to the planet, and he now serves as board chair of Interfaith Power & Light.

Reverend Dr. Durley draws from his experience in organizing and fighting for civil rights. "I have hope because when you've been through a hopeless situation, but stay focused on your goal, then you achieve a victory. You can set another challenge and another victory," he stated.

His involvement in climate action started when actress Jane Fonda introduced him to her stepdaughter, Laura Turner Seydel, to talk about climate change.

"At first, I wasn't convinced. I told them, 'You lost me at polar bears,'" he said. As he started to learn more, he became intrigued. "I saw the connections between public health, redlining, and climate justice. We were not just in a legislative battle, but an ethical and moral one."

Reverend Dr. Durley continued: "Once people understand that climate change impacts them personally, the opportunities to get involved become natural. You can start by pulling tires out of the river like in a Chattahoochee River Cleanup day. Once you get involved, you see you can do even more."

He remembered attending a convening of climate leaders several years ago.

"For two days, there were tons of intelligent presentations about all the data on melting glaciers, heat waves, ocean acidification, and bleaching of corals," he recalled. "At the end of the meeting, the group asked me for feedback. I stood up and said, 'I have a PhD and two master's degrees, but you are the dullest people I've ever met. Here we are fighting for airtime with *The Real Housewives of Atlanta,* and we're sitting around wordsmithing and talking about academic studies. Of course, we need plans and strategic objectives, but we have to be breaking news and impact daily lives to make a difference. That's my observation."

He said you could hear a pin drop. Then came the claps and a standing ovation. He encouraged the group to personalize the climate emergency and make solutions tangible.

This story relates to Reverend Dr. Durley's advice: "We are in a clim-a-demic. There is no vaccine. Prevention is the cure. We have to keep warming to 1.5 degrees Celsius. The climate movement can't be selfish. It must be for the good of all. God created a balanced world. As a nation, are Americans ready to change our lifestyles? Even a little bit?"

Dr. Mark Hyman

Functional Medicine Doctor and Bestselling Author

"Buddhism centers compassion, which is one of the reasons I became focused on healing," said bestselling author Dr. Mark Hyman. He runs the UltraWellness Center in Lenox, Massachusetts, serves as the Head of Strategy and Innovation for the Cleveland Clinic Center for Functional Medicine, and hosts one of the leading health podcasts, *The Doctor's Farmacy*. He's written more than fourteen *New York Times* bestselling books; is an expert on food, health, and the environment; and is a regular contributor to *The View*, *CBS This Morning*, and *Good Morning America*.

As a doctor who practices functional medicine, Mark credits his undergraduate degree in Buddhism from Cornell as his wake-up call for a life in service. "Bodhisattva is not about enlightenment for your own sake, but to help others and relieve suffering. It takes a worldview."

After college, Mark went on a backpacking trip and decided that serving others through healing was his calling. During medical school, functional medicine was referred to as "clinical ecology," which Mark points out is an appropriate name for how human health and the planet are intertwined. His unique educational background, combined with an ability to see the whole person, primed him for a future in functional medicine practice. Mark's work reaches millions of people through his medical practice, books, engaging online platform, popular podcasts, and television appearances.

In his current practice, some of Mark's patients are concerned about climate change and the horrible headlines. He tells them, "Either you can embrace hedonism and do nothing or decide to become an activist. Action helps alleviate some anxiety. How you want to contribute becomes the biggest question."

Mark sees a positive shift in the future as more people "embrace a holistic future and realize there's a lot you can do, individually and politically, to make a difference." He gave this advice on how to get involved in the climate movement: "Take action. Whether it's composting, voting, volunteering, donating, or trying out brands. These small steps make a difference for your personal mental and physical health but also your overall well-being. Empower others with knowledge, support, and encouragement."

When it comes to intergenerational partnerships, Mark recommends talking to your kids as the crucial first step. "As simple as it seems, making service and caring for the planet a conversation around the dinner table helps them become family values. Those conversations are a powerful force to influence others and shift the culture toward global climate solutions," said Mark. "Community is essential to fuel lasting culture change."

WHAT NEXT?

As you check out the strengths, the attributes, the areas for development, and ways to avoid burnout, do you think you are a Sage? Does this profile seem to match with who you are and your relationship to nature? If not, check out the other profile chapters. If yes, ponder your spirituality or faith and your role in the climate movement.

Three spiritual or faith-based lessons I know from my time spent in nature:

1. ____________________
2. ____________________
3. ____________________

Three ways that people could benefit from this spiritual knowledge or connection:

1. ____________________
2. ____________________
3. ____________________

Here's how I would link faith and climate action:

1. ____________________
2. ____________________
3. ____________________

The 72-Hour Check Back

Come back to this section within three days. Think about how you'll connect faith and climate action. Tell a loved one that you're embarking on a daily practice of sustainability. Ask them to be an accountability partner. Tell them you'll show them how to take action.

My accountability partner is: ____________________.

Reflect on your strategies to relate faith or spirituality to climate action. What did you learn? What do you want to know more about? Where can you apply these connections?

____

____

____

SAGE ONE GREEN THING EXAMPLES

Micro Impact

- Meditate or read a favorite spiritual text outside.
- Spend mindful time with a pet.
- Work in a community garden.

Macro Impact

- Encourage your house of worship or faith community to install solar panels to reduce climate emissions and save on electricity.
- Speak about the connection between faith and climate action in your spiritual community.

If you're ready to get started, read through the 21-Day Kickstarter Plan for the Sage. Scan these daily One Green Thing ideas to inspire your plan. You can help shift the culture by applying the Laws of Simplicity & Consistency, Identity, and Amplification. As your daily actions compound, you'll promote comprehensive solutions to the climate crisis for a more sustainable future. The Kickstarter Plan mirrors part 2, "The Seven Areas to Effect Change."

Feel free to either review the other profiles or skip ahead to chapter 10 for a visualization exercise and chapter 11 to check out the tools, including the Eco-Action Plan, Eco-Impact Top Ten, and Joy Tracker. Then flip to part 2 and put your Service Superpower into action.

THE SAGE

21-DAY KICKSTARTER PLAN

	SUNDAY *know you can't go it alone*	MONDAY *think beyond your age*	TUESDAY *see energy in a new light*
WEEK 1	Plan to travel with a friend to a favorite natural area and meditate or pray.	Spend time outside with an older family member or friend for a visit, walk, or hike.	Share about the challenges of global warming in developing countries with your community or congregation.
WEEK 2	Call a friend you haven't talked to in a long time. Ask how they are doing. Make a plan to call them again.	Schedule a family Zoom call to talk about positive stories on social change. Share your thoughts on climate action.	Learn more about climate justice and host a community conversation about energy equity.
WEEK 3	Plan a Volunteer Day at a community environmental organization, camp, or service organization that connects faith, spirit, and service.	Talk to a Gen Zer about connecting to the outdoors and spirituality. Ask them about climate change. Or share your thoughts with someone older.	Inspire and lead your place of worship to consider investing in renewable energy sources for the facility.

WEDNESDAY *understand that you are what you eat*	THURSDAY *protect the source*	FRIDAY *vote with your wallet*	SATURDAY *love your mother (earth)*
Practice mindful eating to reflect on the importance of food as fuel.	Meditate by a body of water—a river, stream, lake, or ocean.	Go to a farmers market and thank a local farmer. Show them gratitude for their support of the community.	Visit a local park to enjoy public lands. While there, meditate and listen for the sounds of nature.
Have a conversation about the importance of plants as food and how our food is grown.	Observe the use of water in your home or neighborhood and list ways to reduce excess usage.	Commit to buying nothing today.	Take fifteen minutes to learn about faith-based environmental organizations, and introduce your community to those that align with your values.
Organize a plant-based potluck in your community to celebrate the power of food and togetherness.	Email, speak, or call on your congregation or community to support strong drinking water regulations and funding for water utilities.	Skip the takeout this week. Be mindful as you prepare your meals, and say a prayer or note of gratitude before you eat.	Go outside for ten minutes. Listen to the birds, check out the squirrels, or hang out with a family pet. Think about what you've learned about compassion from animals.

Chapter 8

THE SPARK

THE SPARK PHILOSOPHY:

"Sure, I'm in."

SPARK ATTRIBUTES AND STRENGTHS

Attributes

The Spark ignites the climate movement by raising their hand to join in. Service isn't always about being in charge, speaking at the podium, raising money, writing a policy paper, or leading a prayer circle. Sometimes the power is in showing up. The Spark Service Superpower happily boosts others by attending events and causes meaningful to friends and family. The Spark plays an essential role in creating a credible, inclusive movement. In a great three-minute TED Talk, Derek Sivers speaks about how important the first follower is to any movement. To illustrate his point, he uses a hilarious video of an outdoor music festival. At first, one lone nut is dancing in front of the crowd. As soon as the first follower (or what I call the "Spark") joins in to dance, others are inspired to get up and dance. Within three minutes, the entire crowd is dancing in front of the band. This is how movements happen, but they don't happen without the person who first says, "Sure, I'm in."[1]

Strengths

Strengths of the Spark include an openness to new experiences, a can-do spirit, a dynamic network of friends, and a belief in others. The Spark understands the significance of showing up in service and the power of being there when a friend needs you. The

Spark is trusting, inquisitive, and joyful. Sometimes this Service Superpower isn't tied to a particular issue and may not be a person who considers themselves an environmentalist.

This profile is generous, believes in the need for social change, and values the role of community in personal health and well-being. They're a fabulous connector of friends, colleagues, and family. Their strong focus on relationships means that they're the glue that holds families, friend groups, and organizations together. Though the Spark isn't necessarily extroverted, they're usually the first one to arrive at a party.

Gretchen Rubin, author of *The Happiness Project*, points out "enthusiasm is an act of social courage."[2] Instead of making fun or scoffing at ideas or opportunities, Sparks embody this social courage of being "all in" without apology. Sometimes it's safer to play it cool or act uninterested, but that's not how Sparks engage with life. They are wholehearted, vulnerable, and earnest as they join in singing a goofy camp song or attending the far-out lecture their friends want them to attend. Sparks are confident, easy-going, and unselfconscious. This strength of character and willingness to "put themselves out there" makes having a Spark in your corner essential for movement building.

As cheerleaders, Sparks are also influencers in their own right. They know what's going on because of their substantial friend networks and are incredibly resourceful when it comes to serving communities. Given their vivacious nature, Sparks move in and out of multiple social circles with ease and get along with people from various and diverse backgrounds. They can find common ground quickly and enjoy learning about other people's stories, goals, and interests.

The Spark's enthusiasm also effortlessly recruits others to the cause. Their helpfulness brings joy to the experience and creates space for others to be themselves, take a chance, experience something new, and connect through service. This Service Superpower profile electrifies the cause and moves the idea forward into action.

SPARK CHALLENGES

This profile's three areas for development are to identify your passion, free your calendar, and embrace your vulnerability.

Identify your passion. Sparks are so supportive of others that sometimes they are roped into hours of activities that may not be specifically aligned with their passion. Plus, without a clear focus, the Spark can flit from one activity to another. They need to spend time asking themselves what they're truly passionate about and really want, which can provide clarity for how they choose to spend their time in service. Then, when a friend or family member asks them to join something, they can be confident in saying, "Sure, I'm in."

Free your calendar. Sparks can become overcommitted because they're invited to many activities and opportunities for engagement by their large and diverse peer group. They want to join everything, so freeing up their calendar to organize themselves, focus on their priorities, and clear their head translates into more time for self-care and reflection. Sparks can feel pulled in multiple directions, and safeguarding downtime is essential to their mental health.

Embrace your vulnerability. Sparks tend to be externally driven and motivated. Their enthusiasm in supporting friends means they're vulnerable in social situations. Acknowledging that vulnerability, even if it's a strength, will help Sparks recharge. Rumination on what they should have said or done sometimes becomes an issue because of their frequent social interactions. While they are easygoing and energetic, showing up for others takes a lot of time and energy, and it's okay for them to reflect on, process, and acknowledge their service.

> **SPARK OVERVIEW**
>
> **Traits:** *enthusiasm, intelligence, dexterity, patience, good humor*
> **Strengths:** *cheerful, helpful, energetic, open, resourceful*
> **Challenges:** *unfocused, overscheduled, vulnerable*

STRATEGIES FOR SPARKS TO RESTORE AND REPLENISH

Sparks can restore by leaning into the Wonk. They can spend time reading up on issues they care about and intentionally focusing on what they value. Here are other specific ideas on how to reenergize:

Read nonfiction. Because the Spark is so supportive of others' causes, it's important that they carve out time to read about topics that interest them. If they grab a book that sounds intriguing, set a timer for at least fifteen minutes, and dig in, the simple act of reading can help them be more in the moment and think less about all the demands on their time.

Spend time alone. The Spark can go for a walk, take in a bit of fresh air, listen to soothing music, and look around. The physical activity and time in nature can lower cortisol, decrease blood pressure, and increase an overall sense of well-being.

Self-reflection. A helpful exercise for the Spark is to set a timer for twenty minutes and pray, meditate, or journal. This time serves to re-center their attention and allows them to tune in to their thoughts, feelings, hopes, dreams, and goals. They can write about their values and passions and also check out the visualization exercise in chapter 10.

Spark Eco-Hero Profiles

Ami Becker Aronson

Philanthropist and Responsible Disruptor

"Some of my earliest memories relate to drought. We were not supposed to let the water run while we were brushing our teeth. We took short showers, not baths," recalled Ami Aronson, president of the Bernstein Family Foundation. "Protecting the environment was a way of life for us; I don't remember making a conscious decision about it." Ami's father was a college professor at University of California at San Francisco and taught environmental occupational health. "The connection between health and the environment was all around me, including in our dinner conversations," she continued.

As a Spark Eco Hero, Ami's work amplifies and invests in sustainability, the arts, and Jewish history and culture, and her leadership centers on empowering others. "Sure, I'm in" is a common refrain as she bolsters her personal mission and her influential and inspiring network of changemakers. Even though she's technically a Philanthropist, Ami pointed out, "It's not just about investing in projects but showing up for your community."

Hope is part of Ami's confident, easygoing spirit. "I witness hope every day in the amazing people I meet. My nephew is headed to college in the fall and is working on technology to help monitor forest fires. My eighty-five-year-old friend recently moved to Israel to invest social capital in health technologies in underserved communities. A dear friend of mine was incarcerated for twenty-two years. As a spoken word and visual artist, his story and kind spirit inspire me. Hope is everywhere if you look for it," she remarked.

Ami understands that the climate crisis looms large. "You can be paralyzed by the enormity of the crisis, or you make a difference. If you are intentional about doing something each day, it'll be enough for your lifetime."

In climate action, Ami recommends that you start local with your impact. "I think so much of getting involved comes from proximity. Our country needs us. Start locally at your school. Attend a fundraiser. Read. Research. Find someone who inspires you."

Ami also insisted that "you don't have to reinvent the wheel" because so many organizations out there are doing amazing work. In true Spark fashion, she suggested: "Join a campaign. And share your joy of action with someone else."

Shane Doyle, EdD

Educational Consultant and Film Producer

"My involvement in the environmental movement was incremental. Ever since I was a kid, I knew we needed a clean earth. I wanted to be on the right side of history preserving our natural resources," stated Spark Eco Hero Dr. Shane Doyle, president of Native Nexus, an educational consulting firm. As a member of the Apsáalooke nation from the Crow Agency, Montana, Shane was glad to help when nonprofits would reach out for a land acknowledgment or an honor song or prayer. "There was an ebb and flow of people reaching out while I was in graduate school. Then I became a spokesperson for several campaigns about public lands. I didn't see myself as an environmental advocate. More and more I became engaged in climate change and made sure that Native voices were heard."

When he was younger, Shane never considered a career in sustainability because the job opportunities seemed unclear. "Growing up on the Crow Indian reservation, I knew I wanted to return after college.

Education was a way to go back home and use my new skills to reinforce the Crow community value of respect for water, air, animals, and nature."

Shane shared, "I am hopeful for our planet because we have more knowledge than we've ever had. We know what we are up against, and the solutions are there. We need to tell young people that we can make a better day. Don't sit back and wait. Quietly go about garnering support. And lead with love."

When asked about Generation Z, he pointed out that this new generation will make an impact on public policies and law and that they're committed to environmental issues. Two of Shane's daughters, for example, are plaintiffs in *Held v. Montana*, the historic youth climate action lawsuit to go to trial and prevail against the state for failing its constitutional duty to guarantee Gen Z the right to clean air and water. Shane also thinks we must remind young people of the positive social changes we've witnessed. "I've seen a sea of change in my lifetime that I would have never thought possible."

With respect to where to start in climate action, Shane emphasized Plains Indian culture: "They lived a ceremonial life for thousands of years. They were tuned into the natural world, our space, community, animals. If we walk like that and are aligned with an environmental ethic, we become role models for those around us."

Shane recommended being intentional about the small decisions we make every day, like how to travel or how much and what we eat. "You don't have to be religious to live life in a ceremonial way. It's a mindset and spirit of divine circumstance."

Colleen Wachob

Co-CEO and Cofounder, mindbodygreen

"We received lots of questions about the 'green' part," observed Spark Eco Hero Colleen Wachob, cofounder of the online wellness brand mindbodygreen, which she created with her husband thirteen years ago. "We knew that a clean environment was the foundation for health and well-being."

An international relations major from Stanford with a decade of marketing and merchandising experience with Fortune 500 companies like Amazon, The Gap, and Walmart, Colleen's open, joyful, and energetic personality thrives on bringing people together and supporting their work. In true Spark fashion, her innovative online platform and community at mindbodygreen embodies this spirit of service and reaches more than fifteen million viewers a month, boasting a community of more than five million followers.

From the latest health trends to environmental advocacy opportunities, her company—whose motto is "You. We. All."—ensures that its readers understand the wellness landscape and critical intersections with planetary health. "The world needs healing, and this motto speaks to how we're all connected and how we want to accelerate a greater understanding of ourselves, our neighbors, and the planet. We try to identify trends before the zeitgeist. For example, we published articles on eco-anxiety and the connection between mental health and the environment years before the United Nations declared 'code red' for humanity."

Science—"Everything from circadian rhythms and getting enough sunlight to grounding your feet into the earth"—confirms that we're hardwired to connect to nature. "I'm not a scientist. But I love shining a bright light on groundbreaking research, trends, and thought leaders." Colleen sees mindbodygreen as "curious curators of content." The goal of the platform is to "give a voice to people and movements that are meaningful."

The news about the climate emergency is grim and scary. "What gives me hope is that we're starting to talk about all of this now. It's a nuanced conversation. It has to be solutions-oriented. We need tools to help people understand the shifts they can make in their everyday lives."

When asked about her favorite solution, Colleen shared that much of the mindbodygreen community has embraced a more plant-based diet. "There's such excitement about plant-based options in a way I haven't seen in a long time. It doesn't have to be all-or-none thinking." To become involved in climate action, Colleen advised that "the most important thing is to find what works for you."

WHAT NEXT?

Now that you've read about the strengths, attributes, areas for development, and ways to avoid burnout, does the Spark Service Superpower resonate with you? If yes, make a list of your strengths and how you think you can use your traits to help the climate movement.

Three strengths I have that could help my community or the movement are:

1. ______________________________
2. ______________________________
3. ______________________________

Three things the climate movement could do a better job of are:

1. ______________________________
2. ______________________________
3. ______________________________

Three areas of action that would be a good match for my talents are:

1. ______________________________
2. ______________________________
3. ______________________________

72-Hour Check Back

Make a promise to yourself to revisit this section within three days. Over the next 72 hours, survey your friends and family and identify those who are already involved in the environmental movement. In the space below, list three people you know who support, volunteer, or work in conservation, energy, food, or sustainability. Ask them to reach out the next time they have an event.

The three people I asked to reach out to me are:

1. ______________________________
2. ______________________________
3. ______________________________

Take a few minutes to think about what you've learned in the past three days about yourself, your community, and climate action. What do you want to know more about? What areas are you most interested in supporting? How can you best apply your strengths to climate action in your community?

____

____

____

____

____

____

SPARK ONE GREEN THING EXAMPLES

Micro Impact

- Say yes when asked to attend a climate change lecture.
- Bring reusables to a friend's cookout.
- Join a friend's fundraising event for an environmental justice organization.

Macro Impact

- Volunteer at a local trail cleanup day.
- Support national climate action by writing to your federal legislators or by visiting their offices to ask them to pass comprehensive legislation.

If you're ready to get started, read through the 21-Day Kickstarter Plan for the Spark, which mirrors part 2, "The Seven Areas to Effect Change." As you apply the Laws of Change, you'll see that the Laws of Simplicity & Consistency, Identity, and Amplification mean that your small, daily actions compound to shift the culture and promote comprehensive solutions to the climate crisis.

Feel free to either review the other profiles or skip ahead to chapter 10 for a visualization exercise and chapter 11 to check out the tools, including the Eco-Action Plan, Eco-Impact Top Ten, and Joy Tracker to check in on your mental health. Then flip to part 2 and put your Service Superpower into action.

We are the superheroes the earth needs right now. By showing up in service, embracing compassion, and speaking truth to power, we can become the awesome ancestors that future generations deserve.

THE SPARK

21-DAY KICKSTARTER PLAN

	SUNDAY *know you can't go it alone*	MONDAY *think beyond your age*	TUESDAY *see energy in a new light*
WEEK 1	Try a new movement or exercise class with a friend.	Support a local environmental education program or outdoor camp.	Sign up for renewable energy options through your local power company.
WEEK 2	Sign up for volunteer day in your community: a food bank, school, library, or retirement home—whatever works for you.	Talk to someone in another generation about climate change. What are they concerned about? What solutions most excite them?	Buy carbon offsets for your family but beware of greenwashing.
WEEK 3	Start a conversation to learn more about and share in your friends' passions for faith, self-care, journaling, meditation, or prayer.	Re-tweet a friend's call to action on an environmental issue.	Share a friend's or environmental organization's post on bringing more clean energy options to your area.

WEDNESDAY *understand that you are what you eat*	**THURSDAY** *protect the source*	**FRIDAY** *vote with your wallet*	**SATURDAY** *love your mother (earth)*
Sign up for local composting in your community or encourage your community to start a composting co-op.	Plan a visit to a local body of water with a friend—a lake, river, or ocean.	Try replacing single-use plastic in your home. Make a list of what is easy and what is hard to eliminate.	Plan a picnic with friends.
Try "Meatless Mondays" for a month to focus on plant-based recipes.	Google your watershed to find out where your watershed boundaries are.	Commit to using real stuff in the kitchen: real plates, cloth napkins and towels. Cut back on paper goods.	Learn how to listen for native birds and spend time outside listening for local species. See if you can tune your ears into the local sounds.
Look at food labels when you shop and work to choose the options with the fewest ingredients.	Tell your family about the importance of supporting causes that keep our waters clean and protected.	Follow a friend's advice on where to buy the best sustainable products.	Agree to go on an outdoor adventure when asked by your Adventurer friends.

Chapter 9

THE WONK

THE WONK PHILOSOPHY:

"We can do it."

WONK ATTRIBUTES AND STRENGTHS

Attributes

The Wonk Service Superpower profile centers on a passion for data, scientific studies, charts, and new information. The Wonk also has an unquenchable thirst for knowledge about the planet, wildlife, and humans, and a desire to empower people. They translate climate solutions in a way that helps others understand that a better future is possible.

Strengths

The Wonk is highly analytical and understands that the right technology already exists to solve the climate crisis. Their belief in technology is tempered by the reality that proposals to make Mars habitable and geoengineering (we'll save the planet by projecting stuff into the sky) are likely focused on the wrong direction. Like Neil DeGrasse Tyson, the Wonk believes in protecting this planet first and ensuring its long-term viability for humans and other living things.

In service, the Wonk is a fixer and a problem solver. They can break down big issues into smaller tasks, which means this Service Superpower involves the keen ability to operationalize ideas. Wonks are organized, driven, and task-oriented. They love to make things happen. While not necessarily academic, they understand how to apply their operational understanding to achieve maximum impact. They can quickly identify what's working and what's not in a situation and bolster others' comprehension of what needs to be done next. Wonks are intensely creative in their solutions, and like the 1980s television hero MacGyver, they can use the equivalent of duct tape and bubble gum to solve tough problems.

As much as the innovation around clean energy, technological advances in our grid structure, and conservation strategies inspire the Wonk, this profile understands that climate change isn't a technological problem—it's a people problem. The Wonk's solutions orientation and "get it done" attitude inspire others to move from anxiety to action. Therefore, even though this profile tends to be outcome- and results-driven, they also appreciate that unlocking the creativity of teams and a people-centered approach drives innovation.

In a larger service context, Wonks make the world go round. It's not enough to identify a problem or want to help. Movements must rally around concrete answers to thorny issues. The Wonk's cleverness makes the future more tangible, and that service resonates with others. This Service Superpower can both help express the urgency of the climate problem and raise awareness of the available solutions. The Wonk has an important role to play, both in creating a new future that's more livable and in painting an exciting picture of what's possible. This profile inspires us to work together to create a regenerative economy—a future where we create systems, operations, and interactions that innovate, reuse, recycle, and repair the communities we live in and that the earth depends on.

WONK CHALLENGES

Three areas for this profile to develop are to be more social, avoid using jargon, and learn to relax.

Be more social. As someone who might be a technocrat or policy pro, the Wonk tends to spend a lot of time analyzing data, projecting outcomes, and running scenarios for legislation, algorithms, or markets. It's important for the Wonk to lean in to the Spark every now and then and hang out with friends to take a break, learn from others, and get out of their routine. Spending time with friends can free their mind for a few moments and result in even more creativity.

Avoid using jargon. Most Wonks know how to translate complex ideas into plain language for others to understand, but it's easy to get caught up in the lexicon of a specialty. This profile needs to constantly remind themselves to be mindful of their language when they talk about science, technology, and climate solutions.

Learn to relax. Given that Wonks are so talented at transforming ideas into workable projects, they sometimes become overly task-driven and need to focus on the present moment. Engaging their senses and deep-breathing exercises can restore a sense of calm and keep them centered.

WONK OVERVIEW

Traits: *problem-solving, originality, cleverness, inventiveness, grit*

Strengths: *thorough, precise, curious, assertive, skillful*

Challenges: *introspective, jargony, tense*

STRATEGIES FOR WONKS TO RESTORE AND REPLENISH

People ask a lot of the Wonks. They shoulder a big burden because of their analytic skills, problem-solving prowess, and ability to operationalize ideas. Here are ways for Wonks to restore:

Put the book down. Wonks tend to be workaholics. They also like their content, whether it be scientific, analytic, or technical. As a result, they need to turn off the laptop or put the book down and look up. This may mean setting a timer to get up and move, go outside, play with a pet, or call a friend. The Wonk will benefit from putting work aside for a few minutes each day.

Learn about pop culture. Wonks are tremendously mission-focused. Therefore, they may miss the larger societal and cultural trends happening around them. To communicate climate solutions and actions, understanding pop culture is helpful. The Wonk doesn't have to subscribe to *Us Weekly* to learn more (although my best friend gifted me with a subscription to this magazine because I was so out of it!). They can reach out to younger people they know and ask them what they like, what's cool, and what they care about. Then the Wonk can think about what they learned the next time they're crafting a message or solving a problem.

Chill out and lighten up with friends. Given that Wonks spend a lot of time alone, spending time with friends and family can benefit them. Watch a movie together. Go for a hike. Have dinner together. As simple as it sounds, making time for friends can help restore the soul. Wonks should be sure to schedule these activities and make them happen.

Wonk Eco-Hero Profiles

Gigi Lee Chang

Entrepreneur, Advisor, and Investor

"My journey in service happened over time. I followed my interests, and over time as I followed different professional opportunities, I found that's where I was most fulfilled," said food pioneer and entrepreneur Gigi Lee Chang. Designated one of the top women leaders in the "future of food" by *Forbes* magazine and a Wonk-Influencer, she thrives on transforming research into reality and operationalizing ideas.

"I came from an entrepreneurial family, so my desire to go out on my own wasn't surprising," but it wasn't until Gigi became a mother and started making her own baby food that she realized her path. Her son loved what she made. "When my friends started saying that their kids weren't eating the baby food they bought at the store, I started researching palate development, organic standards, and the role of food in a child's development." Gigi wondered, *Why was the food I was making such a curious thing to other moms and caregivers?* In addition to digging into the literature, she interviewed friends. Many didn't have the time, the know-how, or the energy to make their own baby food. Gigi saw the business opportunity and founded baby food company Plum Organics in 2006. Sourcing to become USDA-certified and meet the standards of distributors like Whole Foods "helped me appreciate how complex our food system is and how challenging it is for consumers."

Now a managing partner at BFY Capital, Gigi uses her Wonk skills to create opportunities for others to grow their vision. Her firm provides capital for companies committed to transparency, clean labels, and sustainability. She recognizes that not everybody can afford or is aware of the benefits of regenerative food. Her work to bring more sustainable brands to scale is to "not let perfection get in the way of progress."

Gigi also underscored that "voting with your dollar is an effective way to get involved in the climate movement. Every bit counts. Don't be overwhelmed. Take on what you can and feel good about it." She recommends understanding a company's mission to see what resonates: "It's not always obvious what will draw you in. There are incredible opportunities to apply your talents in service to support the climate movement in what you are already doing every day."

Dan Puskar

President and CEO, Public Lands Alliance

"I got into advocacy not because of the environment, but because of land," explained Dan Puskar, president and CEO of the nonprofit Public Lands Alliance and a Wonk-Influencer Eco Hero. "I grew up on a three-square-block residential development in rural New Jersey, surrounded by forests and fields and creeks to play in. Every summer my family and I went to the lake district in Maine, but I didn't think about it as 'environmental' stuff."

During his junior year in college, he discovered an exciting opportunity to study abroad in Botswana, but the condition was that he had to take ecology. "I was learning from native peoples living near the Okavango Delta and exploring conservation, anthropology, and ecology." This experience changed his life. "I understood the profound relationship between land, place, people, and biodiversity. The scale of it opened my eyes."

In his role at the Public Lands Alliance, Dan follows the ins and outs of Capitol Hill and tracks how appropriations and proposed federal and state policies will affect local, state, and federal lands and the nonprofits that bolster them. "Public Lands Alliance exists to convene grassroots nonprofits, to share ideas on what works and what doesn't, and to serve as an engine for capacity-building. We also work on big issues in public lands today, like creating safe, inclusive spaces for all people to enjoy the outdoors."

Nonprofit partners to public lands can break down barriers and form opportunities for trust and inclusivity. "I was very lucky early in my public lands career to find a Sierra Club Colorado–organized backpacking trip for the LGBTQIA+ community to the San Juan mountains. It meant something to be able to backpack in a national forest with people like me," Dan recalled. More nonprofit partners are also advocating for more inclusive experiences for Black, Indigenous, and People of Color and other underresourced communities.

With a true Wonk spirit, Dan remarked that "setting aside land for carbon sequestration and supporting soil health is an incredible opportunity to curb carbon emissions. The good news is that a lot of people care about land conservation. We don't need to have the same underlying motivations in order to work together. And there is no action too small. Talk to people who are doing things you're interested in. Local impact matters just as much as international impact. We need all of it."

Dan Wenk

Former Superintendent of Yellowstone National Park

"I never used to think about how my actions today might affect my kids and my grandkids, but I do all the time now," said Dan Wenk, former superintendent of Yellowstone National Park. A Wonk Eco Hero, Dan served in the National Park Service for forty-one years and says that for significant climate policy action to happen, "education is incredibly important because people have to accept there's a problem before they change their behavior."

Dan spent his youth exploring woods and streams around his house. "I wanted to go into the design field, and landscape architecture seemed like the right place to be. That interest led me to the National Park Service [NPS]." The built environment of the national parks, called *parkitecture,* still inspires him. "Within the national parks there are great iconic structures, like the Old Faithful Inn, that have withstood the test of time. They are masterpieces of architecture."

As a bureau within the Department of Interior, "NPS doesn't regulate as much as educate," said Dan. "One of the ways we could educate best was not by hitting people over the head with 'this is what you need to do,' but by showing them great examples of environmentally friendly architecture and landscapes. NPS has an audience of 300 million people a year. You can see how we do things and apply them to your daily life, at your home, or in your community."

When asked about the climate crisis, Dan remarked that "*depressing* is an apt word to describe it. You don't have to look very far to see the impacts of climate change. Yellowstone was established 150 years ago, so we have an incredible amount of data. I was in Yellowstone twice, from 1979 to 1984 and 2011 to 2018. By 2011 there were thirty less days of freezing temperatures a year. Snowpack was affected, which impacts river systems and water storage facilities. We've seen changes in wildlife migration, in fires and duration, and in habitat."

Dan remarked that the climate change trends are "undeniable," but that lawmakers tend to look at "episodes." "The problem is that we change our experiences when climate-induced extreme weather interrupts our life, when we need to change our actions." He urged us all to work toward strong global climate policy, promote climate change education, and support young people in their advocacy.

WHAT NEXT?

Now that you've read the strengths, attributes, areas for development, and ways to avoid burnout, ask yourself if the Wonk Service Superpower resonates with you. What are your strengths? What skills do you take for granted that others may not have? When do you feel most useful?

Three strengths I have as a Wonk are:

1. ______________________________
2. ______________________________
3. ______________________________

Three ways that the climate movement could improve:

1. ______________________________
2. ______________________________
3. ______________________________

After reflecting on my interests and talents, here are three ways I might be able to help:

1. ______________________________
2. ______________________________
3. ______________________________

72-Hour Check Back

Think about what climate action issues you're most interested in. Clean energy? Species extinction? Carbon sinks? Sustainable design? Finance? Write down the three areas you'd like to learn about, and make a commitment to spend fifteen minutes researching each topic in the next 72 hours.

The three topics I will research include:

1. ______________________________
2. ______________________________
3. ______________________________

After the three days pass, write down what you learned in these areas. How did researching these topics make you feel? What surprised you most?

____

____

____

WONK ONE GREEN THING EXAMPLES

Micro Impact

- Attend a lecture on environmental issues.
- Research one of the top five climate solutions.

Macro Impact

- Give a talk about climate solutions at a local community center.
- Write an op-ed in the local paper about endangered species.
- Testify at your next city council meeting on the need for water conservation and drinking water protection funding.

If you're ready to get started, read through the 21-Day Kickstarter Plan for the Wonk. Scan these daily One Green Thing ideas to inspire your plan. You can help shift the culture by applying the Laws of Simplicity & Consistency, Identity, and Amplification. The Kickstarter plan mirrors part 2, "The Seven Areas to Effect Change."

Feel free to either review the other profiles or skip ahead to chapter 10 for a visualization exercise and chapter 11 to check out the tools, including the Eco-Action Plan, Eco-Impact Top Ten, and Joy Tracker to check in on your eco-anxiety and your overall mental health. Then flip to part 2 and put your Service Superpower into action.

We can address eco-anxiety through intentional intergenerational conversation, leaning into our Service Superpowers, and taking action each day to shift the culture to scale the climate market and policy solutions we need.

THE WONK
21-DAY KICKSTARTER PLAN

	SUNDAY *know you can't go it alone*	**MONDAY** *think beyond your age*	**TUESDAY** *see energy in a new light*
WEEK 1	List the most inspiring and effective articles or books you've read on community building.	With other generations, watch a documentary about green living, the climate crisis, or sustainable practices and talk about it afterward.	Take fifteen minutes and find out whether you can switch to clean energy through your utility. If yes, make the switch.
WEEK 2	Research self-compassion, self-care, and how to avoid activism burnout.	Make a plan to read seminal environmental books from different decades (e.g., Rachel Carson, Al Gore, Greta Thunberg, etc.).	Take ten minutes to research the next meeting of your town or city council. Plan to share your thoughts on the importance of sustainability in a letter or in person.
WEEK 3	Learn more about conservation or climate action organizations in your area. Make a plan to show up at their next meeting.	Ask someone from a different generation—family or friend—about positive social changes they've witnessed. Compare notes.	Do an assessment of your home's appliances and make a goal list of energy-efficient options.

WEDNESDAY *understand that you are what you eat*	THURSDAY *protect the source*	FRIDAY *vote with your wallet*	SATURDAY *love your mother (earth)*
Research and share how to support sustainable agriculture in your community.	Take ten minutes to brainstorm ways to reduce water usage in your home, and choose a strategy to implement.	Track your use of single-use plastics for a day and figure out solutions to use less.	Work outside and enjoy the nature around you.
Influence and educate friends on the mind/body connection and the importance of nutrition in brain function.	Read an article about the importance of water conservation and reflect on what you learned.	Google trends in sustainable design. What's most exciting to you? Are there projects in your neighborhood that could benefit?	Sign up for wildlife conservation group's action alerts to support wildlife protection.
Share information about the importance of soil health in the foods we eat.	Attend or watch a lecture about water as a resource and how best to preserve it.	Take five minutes to Google the best sustainable brands for whatever household items you need this week. Then make the switch.	Consider how being outdoors affects how you feel and research the connection between nature and mental health.

Chapter 10

FUTURE 2030: APPLY YOUR SUPERPOWER THROUGH VISUALIZATION

THE TWINKLE IN THEIR EYES

When I was growing up and my grandma would talk about my dad, she'd say, "That's when you were just a twinkle in his eye." She meant that she was talking about a long time ago, way before I was born. When I discussed the mindset of compassion and resilience earlier in the book, I encouraged you to create your own "why" for climate action. My why is all about the twinkles in my eye and my kids' eyes. I want to be an awesome ancestor.

My paternal grandma, Wanda, left high school at seventeen to get married and have kids. When she was in her midforties, she returned to school to get her high school diploma. No one encouraged her, but she did it because she believed in the power of education. My great-grandma on my mother's side, Lovie Jane, didn't know how to read or write. Granted, she was born in the 1880s in Appalachia. Lovie Jane signed her name with an *X*, as my mom recalls. And here I am, writing a book. Let that sink in.

At random moments, I think about my grandparents and the generations of change agents I never knew or met who have helped me. They made advances in science, secured our right to vote, and created educational opportunities. Then there were those whose survival itself was heroic, those who kept going when it seemed impossible. That desire to honor my past and pay it forward to future generations is a strong motivator for me. This intergenerational duty moves across cultures, too, whether it be the concept of the Seven Generations in Native American culture or the Ghanaian concept of "Sankofa," a turning back and reaching for knowledge in the past. In chapter 13 we'll talk more about creating an Intergenerational Partnership, but these ties help ground us as we visualize the future.

We need to act fast. The International Panel on Climate Change says that we have until 2030 to make enormous progress on climate change, or we may reach a point of no return. Even in the polarized political world we currently live in, we must change the culture. Part of that change is individual action, but it's also believing that change is possible.

WHY EQUITY IS CENTRAL TO CLIMATE SOLUTIONS

Equity is a necessary element to addressing the climate crisis. Black, Indigenous, People of Color and low-income communities will be harder hit by climate change extremes. Study after study has shown that these communities experience higher rates of air pollution and associated health effects like asthma and heart disease. High temperatures and air pollution are linked to pregnancy complications, and Black women are hurt the most.[1] Heavy-polluting industrial sites are frequently located near communities of color. When Hurricane Katrina hit New Orleans in 2015, 30 percent of New Orleans residents didn't have cars, which meant it was harder for them to evacuate. And when stranded residents eventually relocated, many did so permanently because they couldn't return to New Orleans without access to transportation.[2]

Historically underserved communities reside in areas with poor infrastructure and lack access to air conditioning, heating, and good insulation, making them more susceptible to weather extremes. It's also typically harder for Black, Indigenous, and People of Color communities to gain access to fire or flood insurance to rebuild after a disaster or to pay for medical bills. Indigenous communities also suffer disparate impacts of toxic pollution, warming climates, and droughts. This inequity is true internationally as well.[3]

As we envision the future, let's clearly articulate how equity plays into this positive worldview and how we can work to protect the most vulnerable. A cultural shift for big climate solutions will take more than a mindset. We need a positive vision to galvanize support. If we can see it, feel it, and believe it, then we can achieve it. That's why, in this chapter, I'm going to ask you to visualize 2030.

YOU ARE YOUR ANCESTORS' WILDEST DREAMS

Before you start visualizing the future, consider this story from a small town in Japan. In 2015, as the town worked on its strategic plan, twenty residents formed a working group to determine what investments the city should make for a better future, as reported by Sigal Samuel of *Vox*. Half of the residents brainstormed needs and investments. The others dressed as if they were from the year 2060, to represent the interest of future generations. The team reached consensus, but only with the "future generations" advocating for significant investments that resulted in the current-day residents forgoing short-term, present-day gains. This role-playing strategy is now used across Japan as part of the Future Design movement.[4] In fact, a whole theory of "longterm-ism" has evolved: obligations and duties to future citizens must balance short-term interests. We owe it to subsequent generations to help them have a better life. In his book *The Good Ancestor*, Roman Krznaric urges us to embrace both long-term thinking and our moral imperative to act to protect future generations.[5]

What do you want to leave the next generation? What type of world do you want them to inherit? How are you going to be a totally awesome ancestor? How are you going to use your Service Superpower to create a better, brighter world? Answering these questions is what this chapter is all about.

STEP 1: LOOKING FORWARD INTO THE PAST—RESPECT THOSE BEFORE YOU

Think back to your grandparents, ancestors you've heard about, or even people you've never met but have read about. What did they do to help you have a better life? Write down the names of three people who created opportunities for you and who helped you. Make sure that at least one of them is someone you've never met. Then write down the action they took to create a better future for you.

STEP 2: WE ARE THE WORLD—NOW SEE IT

What does a world look like in which we value sustainability, have reduced global emissions, and flourish in a regenerative economy? What does it feel like? How does personally valuing climate action and service work? How does equity play a role in the future vision? Envision what a regenerative, positive 2030 would look like. Be creative. Draw it. Write it. Use word imagery. Make it as specific and as hopeful as you can.

Visit www.onegreenthing.org/2030visualization to download the visualization exercise.

STEP 3: YOUR 2030 LIFE

Think about your family and friends in 2030. Where will you live? What will you do for a living? How do you spend your downtime? What does a thriving community look like to you? What do you eat? Where does your food come from? How do you interact with others? What role does technology play? What role do you want it to play? What role does nature play? What role should it play?

STEP 4: YOUR SUPERPOWER IN ACTION

Reflect on how you've used your Service Superpower in the past. Now think about how you might apply it in the future. How can you help create a brighter 2030? How will you use your Service Superpower? What One Green Thing are you already doing? What do you want to try? What areas of climate action do you want to know more about? What do you want to apply at the community level?

STEP 5: BE AN AWESOME ANCESTOR

Now envision a meeting between your current self and your 2030 self. What actions did you take now that resulted in a healthier, greener, more equitable future for your 2030 self? What actions are you most surprised by? What steps are most exciting for you?

Then imagine it's 2050, and you're meeting with a teenage relative. You are the ancestor. What will the young person thank you for? What will the young person wish you had known about? What advice will you give that child of the future? What did you do to help that child have a better future?

CHAPTER TAKEAWAYS

- We need more than a mindset to create culture change for climate solutions: we need a positive vision for the future.
- Climate justice and equity are essential to climate solutions.
- The actions of those in our past, combined with our own actions in the present, impact the future.

JOURNAL PROMPTS

- After completing your visualization exercise, what Service Superpower skills do you want to develop to meet this vision of 2030?
- Think about what habits you might want to change to contribute to the positive vision for 2030. Write down three small things you'll try to do to make a difference.
- What was most exciting about your vision for a regenerative, positive future for the planet? Is there a place you want to visit or hold sacred? Is there a place you want to share with family and friends? What connects you to that geography? How might the experience change as the impacts of climate change become more real?

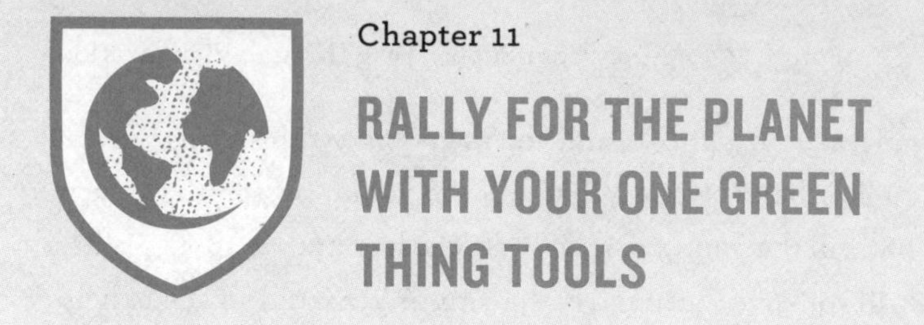

Chapter 11

RALLY FOR THE PLANET WITH YOUR ONE GREEN THING TOOLS

IT'S THE END OF THE WORLD AS WE KNOW IT: THE ECO-ACTION PLAN

I danced around to R.E.M.'s anthem "It's the End of the World as We Know It." I thought it was a funny way to kick off our discussion about our Service Superpower profiles and the Eco-Action Plan. My family, however, did not.

After they took the Service Superpower Assessment, I discovered that my husband, David, was a Wonk; my older daughter, Cady, was a Beacon, and my younger daughter, Susan, was a Spark. As I previously mentioned, I'm a Philanthropist-Wonk composite. Equipped with knowledge about our unique profiles, we sat down to talk about our Service Superpowers and how to create a daily practice of sustainability.[1]

THE ECO-ACTION PLAN

The purpose of the Eco-Action Plan is to create a road map for your daily ritual of sustainability. Tailor it to meet your needs and your Service Superpower. Then share your progress with your friends and the One Green Thing community. Here are the five steps:

Step 1: Planning Time

As you develop your Eco-Action Plan, consider using the journal prompts offered in this book. Ask yourself what's important to you. What do you want to feel about nature, the earth, and the future? What One Green Thing options seem aligned with your Service Superpower?

I sat down with my family to talk about creating a practice of a One Green Thing. We discussed our Service Superpowers, the tools, and the Seven Areas to Effect Change (see part 2). Global climate change was the central theme, and we also discussed energy policy, water quality, species extinction, agriculture, and equity issues. We brainstormed what types of actions seemed to connect with us.

Step 2: Share and Listen

If you're going to engage others with your plan, share and take notes. Think about how your Service Superpowers may differ and where you have common ground.

My older daughter Cady's biggest concern is energy policy. To tackle the root cause of climate change—dirty energy production—she knows we must create fundamental policy changes to shift to renewables. Susan is passionate about educating girls and providing them access to health care and family planning, to promote equity and increase innovation. Educating girls globally is one of the top ten solutions to climate change, according to Project Drawdown. My husband's work reflects his top priority, which is protecting wildlife habitat and public lands. My biggest concern is to ensure that elected officials act on climate. It was fun to talk through how we could apply our different Service Superpowers, work toward solutions, and chip away at the feeling of being overwhelmed.

Step 3: Dive In and Learn

Research the issues you care about and set up a plan for action. Learning about the issues you care about can help guide your daily practice. Check out the resources at the back of this book and the Seven Areas to Effect Change. Each chapter provides suggested One Green Things.

After the initial dinner meeting, Cady started reading about energy policy, and Susan elected to learn more about girls' access to education. My husband decided to double down on his professional work in wildlife management. Our discussion inspired me to study the impacts of a plant-based diet on global warming and review the climate platforms of elected officials.

Step 4: Create a Plan and Track Progress

Create your Eco-Action Plan and write down your priorities and action steps. Research shows that writing down your goals increases the likelihood of success. Recent studies confirm that "performative environmentalism," or individual action, creates the political momentum for policy reform.[2] Check out the trackers in this book to record your actions and how you feel along your journey.

My family and I met again to write down our Eco-Action Plans. Cady is passionate about climate change, so she became a Green New Deal expert and decided to sign online petitions to urge Congress to pass climate change legislation. As a Beacon, she enjoyed posting what she learned to Instagram to try to inspire her friends to take action and speak out for climate action. Susan found it fascinating that equity can help reduce global climate emissions. Given that she is a Spark, access to education deeply resonated with her, and she decided to raise money to support the Malala Fund. My husband, true to his Wonk Service Superpower, continued to read about wildlife management, network with other professionals in the field, and share his knowledge. I decided to try a plant-based diet and invest in and actively support political leaders who will fight for climate action, which aligned with my Philanthropist Service Superpower.

Step 5: Celebrate and Look Ahead

Check in after your first month. What worked? What didn't? What's next? What action do you think had the most impact? What made you personally feel better?

We discovered that action did make us feel better, but it also raised more big questions. Cady felt positive about signing petitions but wasn't sure if they changed minds. (They do.)[3] Susan wanted to do more on equity for girls and didn't know if an eighth-grader making a one-hundred-dollar contribution to the Malala Fund would make a difference. (It does.) My husband felt good about and energized by his professional work. Being intentional with a plant-based diet made me feel better physically. Even though donating money to candidates didn't change federal climate policy immediately, I was glad I took an action each day to move the needle.

We also enjoyed sharing our actions with our friends and knowing we were putting our Service Superpowers to work. In addition to our Eco-Action Plan, we had fun challenging one another to try a One Green Thing each day. From repotting plants to mending clothes, these daily habits create cultural momentum for substantial policy change.

HOW YOUR ONE GREEN THINGS AND SUPERPOWERS ADD UP

This chapter reviews the tools for action to inspire change. Keep track of your daily practice of sustainability. Share it on social media with #onegreenthing and spread the joy. Remember to touch base with your why and your vision for the future as you keep working on your daily practice.

The *Eco-Impact Top Ten* is a way to measure progress. Rather than calculate the specific reduction in your carbon footprint (believe me, it's depressing—like losing one-tenth of a pound every six months), the Eco-Impact Top Ten is a simple list of the activities you can adopt as One Green Things for the biggest impact. Talking about the climate crisis and voting are two of the most significant strategies because we need big, global solutions. These One Green Things do make a positive difference. A new analysis from Project Drawdown and RARE estimate that individual and household actions can contribute to 25 to 30 percent of the carbon reductions we need to abate climate change and keep temperature rise to 1.5 degrees Celsius.[4] If you're interested in an actual carbon footprint calculator, check out the Environmental Protection Agency's calculator at https://www3.epa.gov/carbon-footprint-calculator/.

The *Joy Tracker* helps you assess how your daily sustainability practice impacts your emotional and mental well-being. It also serves as a reminder that being part of the climate movement isn't all doom and gloom. A regenerative lifestyle is hopeful and positive. Refer to your 21-Day Kickstarter Plan and see what activities work best for you. Share what you learn with your friends and family.

In part 2, you'll learn more about the Seven Areas to Effect Change. At the end of each chapter, I've provided an *Action Checklist* so your actions can align with specific areas of change that resonate most with you, your values, and the future you want to create.

CHAPTER TAKEAWAYS

- Create your own Eco-Action Plan as an individual or as a family to start your One Green Thing routine and make it fun.
- Tracking your activities can measure progress. See the Eco-Impact Top Ten to find out the ways to make the biggest impact in climate action.
- The Joy Tracker helps you realize that action can reduce eco-anxiety and stress about the future.
- *Make it fun!* Yes, climate disruption is real, but a daily practice of sustainability can ease anxiety, create community, and bring joy into your life.

For additional resources, please visit www.onegreenthing.org/bookresources.

JOURNAL PROMPTS

- Check out the tools in the appendix. What tool seems like the best fit for you or your lifestyle? What do you want to share with a friend or family member?
- What step of the Eco-Action Plan sounds the most interesting to you? The most fun?
- What are your barriers to following through on an Eco-Action Plan or a daily practice of sustainability? Name three challenges you might face as you implement your plan. Now think of how you might overcome those challenges to stay on track.
- What are you already doing that might work as a One Green Thing? What more do you want to do? What types of action do you think will be most meaningful to you and your family?
- Have you ever tracked your feelings before? How often do you tune in to how you are feeling or sensing? Or how you're connecting with others? What do you think the Joy Tracker might teach you?

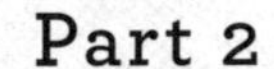

Part 2

THE SEVEN AREAS TO EFFECT CHANGE

About Part 2

Now it's time to apply your Service Superpower. Here I outline the seven areas where you effect change. In each of these chapters, I provide an overview of the problem, potential solutions, and how you can help. You'll also find:

- *Green Touchpoints:* quick check-ins to reflect as you dig into the material
- *Cool Solutions:* promising ideas to address the climate crisis
- *Move from Hope to Action:* an action checklist organized by Service Superpower
- *Journal Prompts:* questions for you to explore further

All the areas of change are vital to climate action, but I've highlighted which chapters will likely appeal to different Service Superpowers. You don't have to read these chapters in order; target those that most pique your interest. The seven areas of change are:

- KNOW YOU CAN'T GO IT ALONE: Philanthropists will connect with this chapter because of its emphasis on compassion as we approach climate solutions.
- THINK BEYOND YOUR AGE: This chapter addresses an intergenerational partnership on climate and will resonate with the Beacons as it describes cathedral thinking.
- SEE ENERGY IN A NEW LIGHT: This chapter walks through specific climate legislation and policies and will likely strike a special chord with Wonks.

- UNDERSTAND THAT YOU ARE WHAT YOU EAT: What we eat and how it's grown affects the climate, which Sparks will appreciate because of the cultural aspects of food.
- PROTECT THE SOURCE: Adventurers will enjoy this chapter given that water is essential both to the spaces they love and in safeguarding public health.
- VOTE WITH YOUR WALLET: Moving the market toward sustainable brands is a critical climate solution, which will align with many Influencers.
- LOVE YOUR MOTHER (EARTH): The focus on preserving biodiversity and how nature increases well-being will appeal to the Sages.

The term *climate action* now encompasses environmental action in general. As you read further, you'll see the interconnectedness of all these issues. Please note that when I refer to carbon emissions I often mean "carbon dioxide equivalent," which is the number of metric tons of carbon dioxide emissions with the same global warming potential as one metric ton of greenhouse gases. I use this shorthand in the text and the *Eco-Impact Top Ten* in the appendix to make the concepts clearer. Part 2 will inform your Eco-Action Plan, support your daily practice of sustainability to ease eco-anxiety and bring more joy into our life, and strengthen your relationship to climate action.

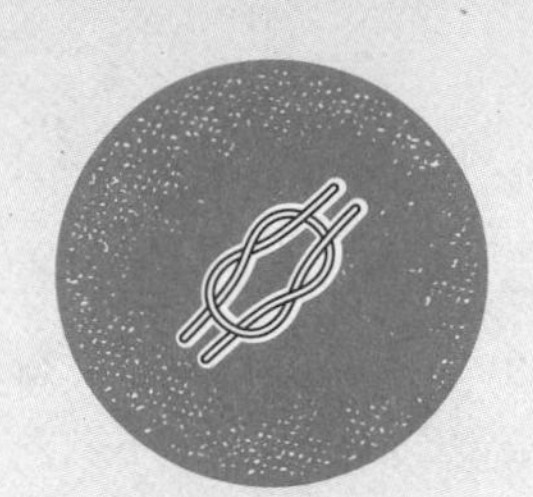

Chapter 12

KNOW YOU CAN'T GO IT ALONE

Refocus on Compassion and Community Connection

My older daughter, Cady, told me not to include the following story in this book. She warned me that it is exactly why people don't want to learn about the climate emergency, that it's too much, too sad, and too overwhelming. This story will likely stay with you for several days, and then you'll see why I told it. Because it's about love and compassion. You'll also understand why we can't go it alone as we face the uncertain future of a warming world.

In September 2020, as cataclysmic wildfires roared across the American West, reporter Capi Lynn of the *Salem Stateman Journal* recounted the story of a family in Marion County, Oregon, outside of Salem, who received fire evacuation orders in the middle of the night. The husband had already left the house to borrow a friend's trailer to gather family belongings. When he returned, police had barricaded the road to his home because the fire had spread. He turned his car around to find a way through to the house, and he saw a woman on the side of the road. She was severely burned and looked like she was naked and barefoot. He pulled over, asked if he could help her, and told her he was looking for his wife. She said, "I am your wife."

Unrecognizable to her husband, the wife had walked three miles in fire and heat so intense that even her clothes and shoes melted. Earlier, when she realized that fire would soon engulf the house and her husband hadn't returned, she instructed their thirteen-year-old son to run for his life. In that instant she also made the heartbreaking decision to leave her mother, a seventy-one-year-old invalid, at home.

After the husband got his wife the necessary medical help, he and rescue teams searched for his son for days. Then they received the news from the authorities. Instead of running away, the son went back to the house and tried to save his grandma. He died in the driver's seat of the family car; his dog was draped over him. Authorities found his grandma's remains in the car with them.

This story is a climate change story.[1]

This is our future.

Despite the heartbreaking losses, this story also illustrates compassion, resilience, and community connection in action. According to GoFundMe, the wife has made a successful recovery. The community raised over $300,000 for the family.[2]

There are countless other examples, like Lytton, British Columbia, where a town had less than twenty minutes to evacuate from catastrophic wildfires, and Elizabeth, New Jersey, where families perished in basement apartments from unexpected flash flooding from Hurricane Ida. These stories serve as cautionary tales about what the devastation from climate change can look like—in our backyards and eventually globally—if we don't act on climate policy. They also remind us of the types of decisions we might have to make when faced with such destruction.

I live in Montana, where fire season is part of summer. But what we're now experiencing is not normal. Fire ecology professor Phil Higuera of the University of Montana recently explained that we're in uncharted territory when it comes to fires in the Northern Rockies. Based on sediment core samples, his team determined that we're in a catastrophic fire cycle in the Northwest, which hasn't happened in two thousand years. No matter the *cause* of these mega-fires in the Western United States—lightning strikes, arson, downed power lines, accidents—extended drought, extreme heat, and "wavy" jet streams from climate change will make fire season more intense and frequent in the future.[3] Indigenous controlled-burning techniques could prove a powerful approach to mitigate these destructive, climate change–fueled fires.[4]

Scientists tell us to expect more flooding with intensification of storms and sea level rise. Rare and powerful rainfall events will dramatically increase during the next century if climate change is not abated.[5] Extreme weather will disrupt our lives in ways most of us can't fathom.

Scientists are encouraging populations to adapt to climate change as we wait for countries to enact global solutions. Meanwhile, fossil fuel companies have spent billions of dollars on slick marketing and lobbying campaigns to cast doubt on whether the climate crisis is even real.[6] To counteract this narrative from Big Oil, we must be ready and willing to share our knowledge with family, friends, and the community at large.

CHANGING HOW WE TALK ABOUT CLIMATE CHANGE

The way we talk about climate change is just as important as what we're saying to people. Sometimes it comes across as "Hi, I'm Heather. I'm an environmentalist, and we're in catastrophic peril. How are you?" That's no way to make friends. And it's certainly not an effective way to rally people around climate action and protecting the environment.

But the reality is scary. According to the Intergovernmental Panel on Climate Change, the United Nations working group of more than 1,400 climate scientists from around the world, we have less than a decade until we cannot reverse the dramatic effects of global warming. And in September 2021, the IPCC's report called for immediate action to curb carbon pollution; the head of the United Nations declared the report "code red for humanity."[7] Every day new studies surface on melting glaciers, catastrophic flooding, rapidly warming oceans, and extreme-weather events around the globe, intensified by global warming.

In the next few years, all of us will have personal experiences with natural disasters and extreme weather that will shake the foundation of who we are and how we relate to each other and the world. Environmentalists like me are trained to use scientific facts and figures to convince people to protect the planet. We assume most people are Wonks, like us, yet the calamitous scientific warnings don't prompt many of us into action. Most of us would rather curl up under a blanket, watch Netflix, and act like everything's fine. Some experience shame when they fully realize what we're leaving future generations. For others, the climate crisis is so enormous it's hard to talk about.

That enormity leads to another problem: systemic desensitization, which means that when the numbers become so large, we can't appreciate the loss. For some, hundreds of thousands of deaths to the COVID-19 pandemic in the United States is incomprehensible. Global losses from wildfires, mudslides, and floods are also on a scale that's hard to process. That's why sharing stories like what happened to the family in Oregon personalizes this crisis and inspires us to engage compassionately with our communities and each other.

We can't go it alone. We need to celebrate positive solutions to climate change that have immediate benefits, even if people don't believe in global warming. Scientists, architects, and engineers agree that technology exists to tackle the problem. Change is possible, if we start talking less about the end of the world and shaming others about how we got here. A less Wonk-like and more Beacon-like approach can paint a vibrant vision of a healthy, regenerative future. We can inspire others to imagine green buildings, rooftop gardens, solar panels, electric cars, and clean energy, water, and air for all.

Part 1 of this book focused on how habits can drive culture change. Right now, there's a lot of animosity and negativity in our culture. It's up to us to break through that pain, highlight what unites us, and focus on compassion and community connection. As Brené Brown tells us, "People are hard to hate close up. Move in."[8] It's time to move in, connect, and work together to create a positive vision for the future. And what's a stronger connector than the planet we all share?

GREEN TOUCHPOINT
Name one reason you picked up this book to learn about your talents in service.

THE POWER OF COMPASSION

Compassion is defined as "the feeling that arises when you are confronted with another's suffering and feel motivated to relieve that suffering."[9] When you feel compassion, your body reacts by increasing the secretion of oxytocin, the same chemical that's released when a couple falls in love or a mother breastfeeds a child. The heart rate slows down, and the oxytocin surge elicits feelings of trust and safety, activating the parts of the brain that are connected with empathy and caregiving.[10] Medical research also shows that stimulation of the vagus nerve, the longest nerve in our central nervous system, can trigger feelings of altruism. The vagus nerve connects the brainstem to the rest of the body by extending from the back of the neck to the throat and into the abdomen. This nerve helps control the parasympathetic part of our nervous system and creates a sense of relaxation because of its connection to breathing, blood pressure, and digestion. One researcher even found that children with high vagus nerve activity are more charitable and compassionate.[11]

Medical researchers have concluded that humans evolved to experience compassion. The evolutionary advantage is that we humans take care of our young, the most vulnerable mammalian offspring.[12] Compassion may also help us attract a mate. A study across thirty-seven countries surveyed more than one thousand participants about what they look for in a mate. Kindness, not resources or attractiveness, was the number-one attribute.[13] As discussed in chapter 1 in relation to the Law of Amplification, positive emotions are more contagious than negative ones.

Published research on compassion and neurology has broad implications:

- Meditation with a focus on compassion stimulates the left hemisphere of the brain and promotes happiness.
- Studies show that a practice of gratitude at the dinner table or in a journal increases a person's overall sense of well-being.
- Time in nature or around "morally inspiring others" increases individual happiness.
- Laughter can help address trauma and promote resilience.[14]

Self-Compassion and Resilience

Self-compassion serves to support overall well-being, address the eco-anxiety from the stressful events in our rapidly changing world, and prevent burnout. Psychologist Kristin Neff's research shows that self-compassion, which is basically treating yourself like you would a good friend, increases resilience. Neff's three elements of self-compassion are (1) self-kindness, (2) common humanity, and (3) mindfulness.[15]

Our inner critic doesn't result in motivation or efficiency. In fact, negative self-talk prevents us from fully appreciating success. Neff urges patients to choose kindness instead of self-criticism, community connection over isolation, and to be present to observe what you're experiencing. Research on the health benefits of self-compassion reveals decreased levels of the stress hormone cortisol and increased heart rate variability, which means the variation in the time between heartbeats. (The higher the variation, the more relaxed you are. You can switch gears from relaxed to action faster.)[16] Self-reflection, meditation, and prayer can all increase feelings of happiness and lead to resilience. That feeling of resilience reduces burnout and energizes you. Additionally, tapping into our inner Sage and connecting with something higher—whatever we choose to call it—can support us and our service to the larger world we live in.

At this point you might be wondering, *If this is a book about service to others and the planet, why are we supposed to think about ourselves?* First, because it's in society's best interests for you to be the best version of yourself. Second, people who are self-compassionate focus on self-improvement to do great work. Finally, Neff distinguishes *narcissism,* an unhealthy obsession and over-estimation of one's importance, from *self-appreciation,* appreciating your strengths and recognizing that all people have goodness. Neff says that focusing on your strengths and talents is a way of "humbly honoring those who have helped us become the person we are today."[17] Taking care of ourselves and focusing on our unique talents both respects our ancestors and links us to future generations, as discussed in chapters 10 and 13.

Reading the news about the climate crisis is anxiety-provoking, and the potential loss is mind-numbing. As your daily practice of sustainability evolves and you become more knowledgeable about the challenges we face, at times you'll need to rest and practice self-compassion. Refer to the section in each Service Superpower profile chapter that suggests concrete ways to rest and replenish. Also make sure to ask for support from family and friends on your journey, including the One Green Thing community.

GREEN TOUCHPOINT

Reflect on who has been compassionate to you in a meaningful way this month. How have you been self-compassionate this week?

WHY WE NEED COMMUNITY

"Know you can't go it alone" celebrates community, not only for social engagement and change but also for our spirit and resilience. I'm defining *community* as broadly as you feel comfortable doing so. Whether it's your local faith organization, an environmental club, an exercise class, a group of high school friends on Facebook, or your neighbors, creating and maintaining a strong community connection is important for well-being, mental health, and yes, even climate emergencies. Make a plan for what happens if you need to contend with fire, snow, or flooding, or how you and your neighbors can help each other safely evacuate together and still access vital necessities, such as food, water, and medications. Get to know people. It's good for your health, and one day your life may depend on it.

Are you ready if you need to evacuate? Check out onegreenthing.org/gokit for instructions on creating a "go kit" so that you're prepared in the event of a natural disaster.

Many psychologists warn that the next pandemic after the coronavirus will be a tsunami of mental health crises, especially for young people. In a 2021 poll during the pandemic, 28 percent of Americans reported increased weight gain (of that 28 percent, almost half gained an average of 29 pounds), 23 percent reported drinking more alcohol, and 31 percent reported problems sleeping. All these factors lead to anxiety, depression, and stress.[18] Concurrently, the climate crisis is intensifying, as extreme weather events increase in duration, scope, and impact around the world. A 2021 global study of more than ten thousand Gen Zers found that nearly half say that eco-anxiety interferes with their daily life.[19]

We need community in order to advocate for climate action and to address the mental health impacts of climate disruption. Community activities increase feelings of trust, safety, and compassion and are positive for our long-term health. Furthermore, global climate solutions, such as educating girls, reducing food waste, and pushing for clean energy require tremendous community action and support.[20] To align with our vision for 2030, we must act together to demand action. Again, this isn't about "saving" anyone or telling people what to do. Instead, it's about being kind, asking questions, reaching out, learning more, and getting involved.

COOL SOLUTION

The Compassionate Schools Project

In a study that has already garnered significant international media coverage, the University of Virginia and Jefferson County schools in Louisville, Kentucky, have embarked on a seven-year program involving more than forty-five schools and twenty thousand children. They created a compassion curriculum that teaches students about mindfulness, breathing, social emotional learning, health, nutrition, and personal development. The study concluded in 2021 but has already provided positive results in student self-esteem, focus, stress reduction, and academic performance.[21]

INTERNATIONAL IMPACTS AND COMPASSION

Countries without significant health infrastructure and access to technology, particularly the poorest countries, will suffer the most from the climate emergency. The World Health Organization predicts that 250,000 to 350,000 people a year will die from climate impacts by 2030 if climate change continues unabated. Malaria, diarrhea, malnutrition, and heat stress will be the biggest causes of death.[22]

The Climate Risk Index analyzes weather data across countries from 1999 to 2018 and advises the United Nations each year on the impact of extreme weather. For the nearly twenty-year time period, the areas most impacted by extreme weather were Puerto Rico, Haiti, and Myanmar. On the other hand, according to the Union of Concerned Scientists, the top country polluters are China, the

United States, India, the Russian Federation, and Japan.[23] As you can see from a quick glance at the lists, the countries who are most at risk from climate-related extreme weather events aren't the biggest polluters. Recognizing the inequity in how climate change hurts those who use the least carbon means we should place justice at the center of climate solutions. International relief efforts will require compassion, understanding, and a commitment to help those most in need.

MOVE FROM HOPE TO ACTION

A recent United Nations report stated "the worst is yet to come" on climate change.[24] Our compassion must extend beyond ourselves and our community. Try not to let the statistics numb you. We're talking about real people and palpable loss.

Think back to the thirteen-year-old son, his dog, and his seventy-one-year-old grandma. Even with unbearable tragedy, there is bravery, love, and compassion. Think of the husband who stopped to help a woman in need during the catastrophic fires only to discover it was his wife. This story is about extending kindness, yet it's also a warning that such a catastrophe can happen to anyone, in an instant.

We need to stand together, move in close, connect, and take action. And yes, we can start small, with one action at a time. We have to start with ourselves, then our community, and then extend outward to the global community. I invite you to apply your daily practice of a One Green Thing to effect change from the bottom up.

PROFILE	REFLECTION	ACTION
ADVENTURER	In what ways can engaging with your community make climate action more joyful?	Visit your local park and bring a journal. Look up. Note signs of the changing seasons. Listen for compassion in snippets of others' conversation or in animal sounds.
BEACON	Did the section on self-compassion resonate with you?	Spend time journaling and reflecting on your vision for 2030. Bev as creative as possible. Share your vision with friends and family.
INFLUENCER	What climate stories or experiences affect you the most?	Invite friends over to eat dinner, play a game, or go for a walk. Talk about your hope for the future and what compassion means to you.
PHILANTHROPIST	How can you bring more family and friends into the movement by personalizing stories and creating in-person community connection?	Donate to or volunteer at a local conservation organization, food bank, community garden, mental health organization, or other cause you care about.
SAGE	What are ways you can talk to your community about climate action?	Check in with yourself. Ask, *Did I get enough water? Have I moved today? How am I feeling?* Then take a drink, get up and move, or call a friend. Practice self-care.
SPARK	How could you create space for Gen Z to share their worries about the future?	Hang out in your yard or a local park and strike up a conversation with a neighbor. You might be surprised whom you meet.
WONK	How can you weave compassion into climate solutions?	Share your knowledge about compassion and its health benefits at your next get-together with friends.

KNOW YOU CAN'T GO IT ALONE ACTION CHECKLIST

Here are ideas for reflection and One Green Things, based on the Service Superpowers, that can help you engage with other people and your community about climate change while practicing self-compassion. To learn more about compassion and the climate movement, visit www.onegreenthing.org/bookresources.

CHAPTER TAKEAWAYS

- Compassion is necessary to help us deal with the emotional challenges of climate change. Compassion is more than empathy—it means we want to alleviate others' pain.
- We're hardwired for compassion, and scientists believe our central nervous system evolved in this way. In addition to protecting our young, compassion is the number-one quality we seek in a companion.
- Compassion doesn't mean "rescuing" someone. It's listening, moving in close, bearing witness, and helping someone on their terms.
- Self-compassion is an important way to deal with the inner critic. As we navigate the immense challenge of global warming, we must also prioritize self-care.
- Community connection increases well-being and health and decreases loneliness. Climate change disruptions mean that we're more dependent on our local communities, infrastructure, and networks. Get to know your neighbors!

JOURNAL PROMPTS

- Reflect on a time when you were a good friend to someone. How did you help them? How did you know you were supporting them?

- Do you know your neighbors? If not, reach out and try to connect. If yes, get to know them better. Ask a question. Then ask a follow-up question.

- When was the last time you went to a community event? What did you learn? Would you do it again? If yes, make a plan to do it!

- Focus on self-compassion. Think of a time when you were hard on yourself and ruminated on a bad decision, mistake, or weird social interaction. Now recall a moment when you were a compassionate, loving friend to yourself. What did you do or not do? How did it feel? Consider taking time each day to make sure you're treating yourself as you would a dear friend. Tell that inner critic to take a hike!

Chapter 13

THINK BEYOND YOUR AGE

Realize an Intergenerational Partnership

LIKE SANDS THROUGH THE HOURGLASS: LEANING IN TO OUR LEGACY

Grandma, turn it up—the A/C unit is so loud I can't hear. I don't want to miss the opening," I yelled to my great-grandma Mary in the early afternoon, right as her soap operas were about to start. I was eight years old, and spending time with her was the highlight of summer vacation. As a special treat, she'd let me watch her "stories" with her. She was partial to *Guiding Light*, but every now and then we watched *Days of Our Lives*. The famous introduction states: "Like sands through the hourglass, so are the days of our lives." Even now it gives me a thrill when I hear the opening. I'm not alone either—the show has pledged to give the iconic hourglass to the Smithsonian's National American History Museum.[1]

I remember how excited I'd get when the show started. What person would come back from the dead? Who had faked a coma? Who had a long-lost identical twin? Imagine my delight when these same themes emerged as I was playing a mystery video game with my younger daughter, Susan. I realized she'd never watched soap operas, so I explained the tropes and we giggled. Like my memories with my great-grandma, I hope one day Susan will recall yelling, "Objection!" with me while playing virtual courtroom dramas. Sometimes seemingly trivial moments can spark delight and remind you of someone you love.

Creating these experiences across generations gives meaning to our personal lives, but they can also positively impact our shared future. Think back to chapter 10, when I asked you to visualize 2030. Then we fast-forwarded to 2050, when you imagined meeting a young person and you're their ancestor.

My great-grandma left me with way more than memories of watching TV. But these uncomplicated moments are filled with love and laughter. They also remind me of poet Mary Oliver's quote: "Tell me, what is it you plan to do with your one wild and precious life?"[2] I want more of that—the feeling of connection across generations. As a society we need to translate this intergenerational hope into policy and marketplace decisions.

EMBRACING CATHEDRAL THINKING AND THE HOURGLASS

I'm inspired by author Roman Krznaric, who writes about "cathedral thinking" and urges us to adopt an ethic of "longterm-ism" as we think of ourselves as ancestors.[3] Our days quickly sift through the hourglass, and therefore we should intentionally think about what we're building for the next generation. He points out that humans can excel at long-term thinking and highlights examples, including the Trans-Siberian Railway, the global eradication of smallpox, the Green

GREEN TOUCHPOINT

Think of an example of "cathedral thinking" that you've experienced or seen in your lifetime—a building, a technology, an ethic, a cultural change, etc.

Belt Movement in Africa, the US Constitution, and Yellowstone National Park.

Past generations thought about us—our needs for transportation, health care, and even wild places. Of course, they missed the mark a lot too, but these incredible advances benefit us all. In climate action, we must adopt this cathedral thinking and start acting in future generations' best interests.

Taking time to consider our personal and generational legacy is a worthwhile exercise. Thinking about life after you leave this planet is what Krznaric calls the "death nudge," a reminder of your own mortality. I prefer to call this phenomenon the "hourglass," not only as an homage to *Days of Our Lives* but also as a way to express the notion of coming to terms with the inevitable passage of time. Behavioral psychology research shows that when people become more aware of their mortality, they're more likely to act in future generations' interests. One study showed that participants who wrote a brief essay about their legacy and how they wanted to be remembered donated 45 percent more to charity than those who didn't write the essay.[4] Another study examined lawyers who first told clients that most people leave money to charity in their wills and then asked the clients if they'd like to donate after they died. This resulted in a 17 percent giving rate, nearly three times the average.[5] This research means that talking about your legacy, thinking about what cathedrals you'll help build, and

GREEN TOUCHPOINT

Think about your Service Superpower and how you've shown up for service to others in the past. What three adjectives do you hope will describe your legacy?

engaging with the next generation can inspire change. Focusing on the hourglass can result in long-term thinking. We need to stand up for Generation Z—they're rightfully worried about the future they're inheriting from us.

LISTENING TO GEN Z ABOUT ECO-ANXIETY

As I mentioned in the introduction, a dinner table conversation with my kids prompted me to dive into the research about the mental health impacts of the climate crisis. My big mom energy offering to *drive* my teen to a climate protest to avoid getting drenched in a thunderstorm resulted in a complete "Mom Fail." My kids' reaction was a wakeup call for me that this entire generation feels alone in climate action. My professional experience in environmental advocacy wasn't fully aligned with our family dinner conversations and daily habits. Furthermore, I realized that we hadn't discussed climate solutions and hope in an intentional conversation with our kids.

Gen Z (born after 1997) and Gen Alpha (born after 2010) both hold climate as their number societal one issue.[6] Despite significant progress with global treaties and the Bipartisan Infrastructure Act, grown-ups haven't exactly modeled the best behavior to take young people's interests in consideration as we develop laws and policies. The political rancor and divisiveness in American politics have impacted our kids' view of the future.

Opting Out of Parenthood

In a September 2021 international survey, one in four young people (ages sixteen to twenty-five) said that they likely won't have children of their own because of their worry about the climate crisis.[7] This statistic stopped me in my tracks. The Canadian documentary, *The Climate Baby Dilemma*, explores this phenomenon. Many parents are having this conversation with their adult children. To be clear, this reluctance for Gen Z and younger Millennials to have kids isn't driven by the media or the science of climate change, it's from their lived experience of extreme weather. It's now commonplace to see wildfire-fueled orange skies serve as a stark backdrop to city skylines around the globe. In the summer of 2023, for example, more than 120 million people in the United States were under threat from air pollution warnings–not from industrial pollution but from pollution called "particulate matter" released from wildfires across Canada.[8] The skylines of major cities from New York to Minneapolis turned a Mars-like orange. 2023 was a record year for disasters, topping nearly $1 trillion in damages in the United States alone. We've seen a five-fold increase in global climate related disasters over the past 50 years.[9]

From FOMO to FODO

In Chapter 1 of this book, I talk about the term that my daughter Cady and I coined for the experience of eco-anxiety. My anxious teen handed over her phone late one night then remarked, "Mom, I'm not on social media because of FOMO [the fear of missing out]. My fear is the impermanence of human existence." The new acronym for Gen Z's constant screen use during the crises of 2020: FODO, the "fear of [humans] dying out." This is reality for Generation Z. They acutely feel the sands through the hourglass because of climate anxiety. It's not good.

TACKLING THE ECO-ANXIETY TRIFECTA

As previously mentioned, for young people, "eco-anxiety" has three aspects. First, *children are suffering from generalized anxiety in greater numbers*. Each child experiences anxiety differently, but the statistics are alarming. The National Institutes of Health indicates that 30 percent of American teens suffer from anxiety. Rates of teen anxiety, depression, and suicide have dramatically increased since 2011.[10] In fact, the United States surgeon general issued a public health advisory on the issue of teen depression and anxiety in 2021. This advisory mentions that climate anxiety is a contributing factor to teen mental health.[11]

Second, *Gen Z is the loneliest generation*. More screen time and less in-person interaction mean a sense of isolation for them, even before and now after the pandemic. Chronic loneliness can be as damaging to a person's health as smoking fifteen cigarettes a day, which prompted the United Kingdom to appoint a loneliness minister in 2018.[12] In the 2022 Cigna Loneliness Index, eight in ten Gen Zers experienced loneliness compared to five in ten Baby Boomers. Today's young people are lonelier than the elderly.

Third is what Cady and I termed FODO, the *hyperawareness of the climate crisis*. Gen Z is asking what the future will look like, where they will live, and what their experience will be like on this planet. This hyperawareness is driven by social media. Young people are witnessing firsthand accounts twenty-four hours a day, seven days a week of droughts, wildfires, flooding, hurricanes, and other extreme weather events by their global peers. A 2022 Youth Climate Survey conducted by Blue Cross Blue Shield of California found that from anxiety to headaches, "four out of five Gen-Z youth have experienced at least one mental-health related impact due to reading, seeing, or hearing climate change-related news."[13] The climate disasters we've gone through and the inequities around the climate crisis are a lot to bear.

Parents often go to extraordinary lengths to help children avoid discomfort (like offering to pick them up in the rain and drive them to a climate rally) to ease our own worry about our children's pain. Recent research shows that kids have to exercise agency in dealing with their anxiety. Swooping in and trying to fix it for them doesn't help.[14]

Like my daughters, the majority of Gen Zers are worried about climate change and the uncertain future they face. To tackle the eco-anxiety trifecta, we need to focus on *connection and compassion,* as I outline in Chapter 12. We can't simply say, it's all going to be okay. That's called the "Reassurance Trap," and it doesn't work.[15]

Instead, ask the young people you love about the climate crisis and the future they're inheriting from us. Really listen. Let them know that they are not alone in climate action. Let them "feel all the feels," and hold space. If climate anxiety is interfering with their daily life, reach out to a mental health professional. Pledge to take action together in real life or "IRL." Sign up for social media about positive climate news and solution. Check out the Kickstarter Plans as an idea to jumpstart your collaboration together. We can address eco-anxiety through intentional intergenerational conversation, leaning into our Service Superpowers, and taking action each

GREEN TOUCHPOINT

Do you know a Gen Zer with anxiety? If not, have you experienced anxiety? If yes, have you talked with the young person about their anxiety? Have you talked to them about global warming?

day to shift the culture to scale the climate market and policy solutions we need.

This generational stress forces us all to think hard about what we're leaving our children, but this situation isn't new. Consider the anxiety of young people during World War I, World War II, Vietnam, and the nuclear arms race of the Cold War. That said, it's time for some serious self-reflection on how we can implement cathedral thinking and build a healthier, more sustainable world as we protect the planet and future generations. Older generations need to share their experiences so Generation Z can see examples of hope and progress. And we need to take action to light the path forward. Yes, action can abate anxiety. Despite the enormity of the issue, we start small and then build momentum.

COOL SOLUTION
Climate Conversations

Create a space for conversation about climate action with your family or friends. Get an intergenerational group to take the Service Superpower assessment and talk about your strengths. Research solutions. Consider what One Green Things resonate with you. Talk about how you can take action together to create a healthier, more sustainable world. Consider collaborating on an Eco-Action Plan (see chapter II).

HOW TO MOVE FROM HOPE TO ACTION

One Green Thing urges a multigenerational partnership to create lasting change. Our experiences and anxiety about climate change may differ, but our need for each other and for action are constant. An Intergenerational Partnership sounds great, but how do we begin? In some cultures, intergenerational conversation and activities are the norm. Here I'm encouraging you to have an intentional discussion about *climate change* and the future.

This initial step requires you to spend time together either virtually or in person. This partnership doesn't have to be with your family; it can be with your community. Meet with someone who's older and someone who's younger, to chat about cathedral thinking and a vibrant, sustainable future. What can you design or lay the groundwork for now that benefits the future? What behaviors can you change individually or together to support momentum for this vision?

As you collaborate, here are some Intergenerational Partnership principles to consider:

- **Ask.** The first step of an Intergenerational Partnership is to ask. Use open-ended questions about the issue—in this instance, climate change—and learn.
- **Listen.** This is a challenge for most of us. Consider setting a time limit and establishing respectful ground rules. Put your phone down. Don't interrupt or think about what you want to say next. For many young people, it's rare to be heard. Pause for two seconds before you respond. Be curious. Ask a follow-up question, and then ask another.
- **Share.** Talk about your own feelings, concerns, ideas. Don't assume the person you're talking with knows the historic event you're referring to or understands your perspective. Walk them through the experience you want to share.

- **Learn.** Commit to growth. This isn't easy, especially as we age. Hearing someone from a different generation explain their fears, dreams, and lessons learned can change your perspective.
- **Laugh.** This is easier said than done when you're talking about the climate crisis. If you're able to incorporate humor, even if it's laughing at yourself, you can open up space to be creative. As comedian John Cleese said, "Laughter connects you with people. It's almost impossible to maintain any kind of distance or any sense of social hierarchy when you're just howling with laughter. Laughter is a force for democracy."[16] It's also a way to break down intergenerational barriers and see the possibilities before you.
- **Brainstorm.** A partnership is not only listening and validating the other person's experience but also trying to form a positive vision of the future together. Ask "what if?" Consider trying out the visualization of 2030 exercise in chapter 10 together. What needs to happen for your vision to become a reality? What can you do now to make an impact for the future?
- **Act.** After the previous six steps, come up with a plan. It doesn't have to be fancy, and you're not going to be graded on it. For example, you can choose a few One Green Things to do together. Promise to share articles on social media. Watch a documentary and then hop on Zoom to discuss it afterward. Or read chapter 14, "See Energy in a New Light," and think about creating an Eco-Action Plan on clean energy. Share your progress with each other.

Check out www.onegreenthing.org/bookresources for discussion guides and other ways to start the conversation.

2050: WHAT WILL WE TELL THE NEXT GENERATION?

In chapter 10 we went through a visualization process and talked about picturing yourself as an ancestor. In this chapter, I'd like to take this concept a bit further. If we want a true Intergenerational Partnership, then we need to adopt the principles and also have a clear vision of what's possible. Given the scientific data of climate change, the urgency can't be underestimated. So why are we talking about thirty years from now? Because systems change takes a long time. Environmentalists have been talking about a World War II–style mobilization effort around climate for years to try to get us on the right track. But can we do it again?

Yes, we can. We know we can, based on the first few months of the COVID-19 response. Recall that the early stages of the pandemic taught us that all our lives and livelihoods are intertwined. Without a doubt, we're globally connected on a biological level. The pandemic also highlighted the power of collective action. The "flatten the curve" response in March 2020 demonstrated remarkable community sacrifice. Medical workers risked their lives to manage the wave of cases and thousands of deaths. Restaurants fed families in crisis and provided meals to emergency responders. Companies pivoted to manufacture face shields, masks, and hospital gowns. These compassionate pivots demonstrate society's ability to mobilize quickly, which is also necessary to address climate change. During the pandemic, climate emissions dropped dramatically. Wildlife returned to metropolitan areas. For several months, nature appreciated our time-out.

No matter your view on why the national COVID-19 response all fell apart, we showed we could pivot fast globally—at least for a few months. As much as we need policy solutions, politics impedes climate action. Our constant election cycle jockeying has led to tunnel vision for our political leaders and increased the division in our country, right when we need to come together the most, to face this crisis.

As a recount attorney for Vice President Al Gore in 2000, I spent thirty of the thirty-six days of the recount in Florida and was in eight different counties. Back then, I thought the rancor was at an all-time high; I couldn't then imagine what we've experienced in the past few years. Given the existential challenges we face, we need to unite at the community, state, national, and global level.

MOVE FROM HOPE TO ACTION

Whatever your political affiliation is, my request to you is to move in close. Connect in your community, have conversations about climate change with your family and friends, and talk about and with future generations and our duty to them. Commit to working toward an Intergenerational Partnership. We can even have fun as we brainstorm together about the cathedrals we can build. Embracing long-term thinking and experiencing moments of delight together will enable us to move forward with joy as we work toward a greener, healthier, more equitable future.

THINK BEYOND YOUR AGE ACTION CHECKLIST

Here are ideas for reflection and One Green Things based on the Service Superpowers, which can help you connect with people across generations so that together you can envision and move toward a greener future. To learn more about intergenerational partnerships, visit www.onegreenthing.org/bookresources.

PROFILE	REFLECTION	ACTION
ADVENTURER	Have you heard of eco-anxiety before or know anyone who has experienced it?	Give a climate presentation at a youth service organization or at a retirement community. Host a panel discussion and bring people of different generations together.
BEACON	What was the 2030 visualization exercise like for you? Did anything in your vision surprise you?	Consider hosting a family Zoom to talk about climate action, climate solutions, and how small actions can make a difference in shifting the culture.
INFLUENCER	What do you think about the impact of technology on Gen Z's concerns about the future?	Try out an intergenerational book club, where you trade off suggesting books that influenced you when you were a teenager. Or try this with documentaries. Hop online to share your perspectives.
PHILANTHROPIST	Take a close look at the research on the biological expression of compassion. Do you notice any of these feelings or sensations when you are in service to others?	Support a youth-led organization and host a roundtable or panel discussion to bring people of different generations together.
SAGE	Apply the cathedral thinking section of this chapter in your weekly reflection or meditation.	Ask a family member or friend who is twenty years older or younger than you about their thoughts on climate change, the future, and what gives them hope.
SPARK	How could you create space for Gen Z to share their worries about the future?	Ask your parents or grandparents to talk about big, positive changes they've seen in society. Also ask them to share personal experiences of social progress. Share your perspective on the world now.

ECO-ANXIETY

WONK 	Think through research on the mental health impacts of climate change and contemplate the policy options to support funding research or climate solutions that might help.	Foster discussion and learning across generations by asking friends or family to share a news article at the dinner table or on a Zoom call.

CHAPTER TAKEAWAYS

- Intergenerational Partnership is a powerful tool for us to create a long-term vision for a more hopeful world.
- Embrace cathedral thinking. What do you want to build over centuries to help create a greener, healthier, more equitable world?
- Remember the hourglass. Reflecting on the passage of time and thinking about the legacy you want to leave behind means you're more likely to consider future generations in decision-making.
- *Eco-anxiety,* defined as "the chronic fear of environmental doom," is a growing concern for Gen Z. The "Eco-Anxiety Trifecta" is Gen Z's generalized anxiety, the loneliness epidemic, and the hyperawareness of the climate crisis.
- Contemplating our legacies leads to a more long-term view in personal, community, and national decisions.
- The Seven Principles for an Intergenerational Partnership are: Ask, Listen, Share, Learn, Laugh, Brainstorm, and Act. To apply these principles, we need to spend time together—either virtually or in person—and talk and listen.
- Working together across generations can shift the political will to adopt market and climate policy solutions on a national and international scale.

JOURNAL PROMPTS

- Think about big social and cultural changes you've witnessed in your lifetime that give you hope for the future. Take five minutes to write down your thoughts.
- What do you want the next generation to know about you and your experience on this planet? How do you want to be remembered? What do you want your legacy to be?
- Write about an important life lesson you learned from your grandparent or an older mentor, or write about a life lesson you learned from someone younger than you.

Chapter 14

SEE ENERGY IN A NEW LIGHT

Recognize That Global Warming Threatens Community Health and Wealth

Hurricane Katrina pummeled New Orleans on August 29, 2005. I remember getting up in the middle of the night, nursing my six-week-old newborn, and watching CNN on mute. I worried that the anguishing images of the Superdome, of citizens stranded on overpasses and trapped on the roofs of homes and apartment buildings foreshadowed our future. I held my baby and wondered what we were leaving her generation. Environmental advocates around the country warned that more extreme weather was on its way if we didn't act on climate as a society. Spoiler alert: we didn't.

Almost fifteen years later, the dayglow orange skyline of San Francisco shocked the world. Journalists covered residents' complaints of smoke alarms going off inside their homes because of the terrible air quality outside. The summer of 2021 brought floods in the subways of New York City and Zhengzhou, China, extreme flooding in Nigeria, Uganda, and Germany, heat waves in India, mudslides in Mumbai, and devastating forest fires in Siberia.[1] My pandemic masks served a dual purpose because they also protected me from coughing fits due to the haze in Bozeman from the fires. Then, on Hurricane Katrina's sixteenth anniversary, Hurricane Ida made landfall in Louisiana, resulting in loss of life as far away as New York and New Jersey. The climate crisis moved from threat to reality.

Now it's time to pour yourself a cup of coffee or tea and channel your inner Wonk. For Beacons and Influencers, this chapter may inspire you and put you on a path to work toward national climate legislation. For Adventurers, Philanthropists, Sages, and Sparks, you might find the chapter dense, but hang in there! I'll provide a basic overview of the science of climate change and climate action. We'll discuss climate justice, too, as we address ways to lift up communities most impacted by climate disruption.

CLIMATE CHANGE 101:
WHAT IS CLIMATE CHANGE?

Climate change technically refers to changes in temperature and weather patterns over a long-term period. *Global warming* is the increase in temperature of the earth's atmosphere from the greenhouse effect, where the sun's rays continue to heat the earth as heavy greenhouse gases act as a blanket and trap other lighter gases from escaping. This phenomenon leads to a warming of the earth, such as what happens in a typical greenhouse. The major greenhouse gases are carbon dioxide, methane, nitrous oxide, water vapor, and chlorofluorocarbons The terms *global warming* and *climate change* are now often used synonymously, as I use them in this book.[2] Here are three key facts regarding climate change:

- According to NASA, humans have increased carbon dioxide in the atmosphere by 47 percent since the Industrial Revolution.[3]

- There is greater than 99 percent scientific consensus that human activity—mostly from energy choices that burn oil, coal, and gas and emit carbon dioxide—is the cause of the climate emergency.[4]
- On August 9, 2021, the Intergovernmental Panel on Climate Change's report, which was approved by 195 countries and based on 14,000 studies, concluded that the evidence that human activity causes global warming is "unequivocal."[5]

Even though 72 percent of adult Americans have accepted the science that climate change is real and a global threat, that number falls short of the 99 percent of scientists who accept this reality. And only 57 percent of adult Americans believe that humans are causing global warming.[6] Compare this to 80 percent of Gen Zers, who believe that climate change is "a major threat to human life as we know it."[7]

Persuading Skeptics

Here are a few facts to share with family or friends who remain skeptical about what we're experiencing. Even though there have been climate cooling and warming periods throughout the earth's history, climate change is "happening 20 to 100 times faster than the most rapid changes in climate history."[8] And while weather may fluctuate from day to day or area to area, climate refers to a longer time period. On average, our climate is on track to be three degrees Celsius hotter by 2100, and over the past century, it's already warmed by one degree Celsius. The largest amount of carbon pollution comes from the energy and transportation sectors. We know fossil fuels are the major contributor because their carbon pollution has a unique carbon dioxide fingerprint, called delta C thirteen, which has dramatically increased since the 1880s.[9]

If some of your family members or climate-curious friends don't believe the international climate scientists, then they might believe the oil and gas industry's own experts. As early as 1959, internal documents show that the American Petroleum Institute (API) stated that oil and gas drilling contributed to global warming. Another example is a 1980 internal document, which details a meeting with oil company members of API and John Laurmann of Stanford University, whom API hired to advise them on the latest in atmospheric research. Laurmann warned that if the world didn't shift to clean energy, then the following would happen:[10]

- 1°C rise (2005): Barely noticeable
- 2.5°C rise (2038): Major economic consequences, strong regional dependence
- 5°C rise (2067): Globally catastrophic effects

Check out Ben Franta's TED Talk from July 2021, where he explains that this scientist told API in 1980 that a warming of 2.5 degrees Celsius could "bring world economic growth to a halt" and suggested that "avoiding the predicted outcomes would require prompt action, since the adoption of nonfossil-fuel energy sources would likely require decades to accomplish."[11] The award-winning exposé from *Inside Climate News* also leaves no doubt that, similar to the tobacco companies and asbestos manufacturers, Big Oil knew its products were the main contributors to global warming and that the impacts would be potentially catastrophic.[12]

CLIMATE GUILT

Do you feel like you should be doing more to help the climate? Or that you're using too much plastic? Not recycling the right stuff? Don't have a hybrid vehicle yet? (Neither do I! I do walk a lot, but I still have a gas guzzler. The market is shifting, and more hybrid and electric transportation options are coming online each day.) Be kind to yourself and remember that you're a product of a system that shifted disposal and energy emissions to you, the consumer, instead of the manufacturers who made all this stuff. The daily ritual of One Green Thing is about intention and cultural change, not blame or shame. Climate activism isn't about perfection—it's about intention, personal growth, and daily action.

FOCUSING ON CLIMATE JUSTICE CAN CREATE HEALTH AND WEALTH IN COMMUNITIES

GREEN TOUCHPOINT

Even with the scientific consensus and extreme global weather events, why do you think only 72 percent of Americans believe in global warming and only 57 percent believe that human activity causes it? In your opinion, what needs to change?

Before I outline clean energy policies to address carbon emissions, let's consider the fact that climate change disproportionately impacts BIPOC communities. Fossil fuel factories are often located there, and Black Americans are 75 percent more likely to live near facilities that produce hazardous waste and air pollution.[13] Research of real estate records and green space show that Black, Indigenous, and People of Color communities have less access to carbon sinks, natural areas like forest, oceans, or soil that can absorb carbon dioxide, and cooling areas like trees and parks; it's hotter in nonwhite communities. Also, air pollution contaminates BIPOC communities more than white neighborhoods, leading to poorer health.[14] Indigenous people, whose ancestral lands consist of the national parks and public lands that environmentalists now revere, also unfairly suffer from climate change impacts. They experience raging wildfires on forests they manage, limited access to water in droughts, and decimation of wildlife that are crucial to traditional diets and cultural practices.[15] This is why we must center compassion, equity, and justice as we tackle the climate crisis.

WHAT IS A CLEAN ENERGY FUTURE AND HOW DO WE GET THERE?

We can fix the climate crisis. First, the technology to curb greenhouse gas emissions exists right now. Second, there are quick fixes that could make big impacts in the next decade. And third, as we discussed in chapter 12, the early stages of the COVID-19 lockdown showed us that the world can mobilize fast. But we need the political will to demand action. That happens when people like you and me stand up and push for change in our communities, state, and nation, and eventually, globally.

Technology

The technology exists for more than *a 70 percent reduction* of current carbon emissions in the next ten years.[16] Renewable energy is becoming cheaper and more available. Electric cars and trucks, wind farms, solar arrays on houses, and energy-efficient appliances, operations, and systems are becoming more commonplace. The

United Nations also has put a strong emphasis on "nature-based solutions" like soil conservation, sustainable agriculture, and ecosystem restoration to store more carbon on the planet. The United States has launched a "30 by 30" initiative to conserve 30 percent of America's natural places by 2030, to create carbon sinks and preserve biodiversity. We can do this.

Ready Solutions

Additionally, existing technology could help us move quickly in the United States. The think tank Rewiring America concludes that modernizing and restructuring the electrical grid of the United States in the next fifteen years could rapidly decarbonize our economy and also create twenty-five million jobs.[17] Re-electrifying the country with five already proven technologies could get us to net zero by 2035. Net zero means reaching a point where greenhouse gas emissions equals the amount removed in the atmosphere, so there is no net increase of carbon emissions. Here's how:

- Bring more wind and solar power plants online
- Transition to electric vehicles
- Incentivize rooftop solar
- Promote energy-efficient heat pumps instead of gas or oil furnaces
- Invest in battery technology to store clean power

With these same technologies, another report from University of California Berkeley researchers determined that 90 percent of the nation's power could be carbon-free by 2035 without passing costs on to consumers.[18] The main point of all this information is to encourage you and show you that we can solve the climate crisis.

Mobilization

Think back to the early stages of the pandemic response, and you'll recall the immense potential for society to act urgently to save lives, as we discussed in chapter 12. Overall daily greenhouse gas emissions dropped 17 to 25 percent during the first months of the coronavirus stay-at-home orders. According to the International Energy Agency, overall global carbon emissions dropped 6 percent in 2020, the largest drop in the post–World War II economy.[19]

That initial pandemic experience showed that if we have the political will and can sustain our effort, we can create a clean energy future. The political will depends on cultural change, which depends on all of us. Then we'll need to rethink how we design products and processes, promote clean energy, grow food sustainably and reduce waste, and preserve clean water and clean air.

One organization who is paving the way is Project Drawdown, which gathered the world's preeminent climate change scientists to identify the top ways to address climate change. They include reduction in global food waste, increased global equity for girls and girls' access to education, more plant-based diets, reforestation, and a new clean energy paradigm.[20] Collective action through policy change is the ultimate solution for global warming.

Why We Must Shift to Clean Energy

One hundred companies are responsible for 71 percent of the world's global carbon emissions since 1988. A closer examination reveals that twenty companies have contributed 35 percent of greenhouse gas emissions since 1965. Familiar names in this top twenty list include ExxonMobil, Shell, BP, Crown, Chevron, ConocoPhillips, and Peabody Energy. The Climate Accountability Initiative reports that these carbon emissions come from use of these companies' products—including gasoline, oil, jet fuel, natural gas, and coal—and from extracting and refining fossil fuels.[21] This analysis demonstrates that moving toward a clean energy economy is a make-or-break proposition to save the planet: we must end our addiction to fossil fuels.

Please note those in the finance world, including banks, investment firms, and insurance companies, have moved the climate action needle significantly. They're already considering climate change in their future scenario strategies, investments, and risk planning.

The good news is that the renewable energy industry is growing rapidly. Solar generation is the fastest growth market in renewables. The Biden Administration aims for solar to provide nearly 50 percent of the country's electricity needs by 2050.[22] Climate Nexus noted that available jobs in the clean energy sector grew 70 percent more than in the overall economy.[23] In 2019, the Bureau of Labor Statistics reported that solar and wind energy jobs grew faster than any other category of job in the country.

Keep in mind that clean energy is not an even playing field. As taxpayers, we need to stop propping up the old technology of oil, gas, and coal, which receive billions in federal subsidies and tax incentives. They have been able to lease federal lands at submarket values for drilling and exploration, release air contaminants like methane with little oversight, and been bolstered by tax credits too. The International Energy Agency reports that comprehensive government policies are essential for renewable energy because the glut of oil production combined with these oil and gas subsidies

GREEN TOUCHPOINT

How do you think we can bring more people into the climate movement?

means that the market alone will not force the necessary reform to curb climate change.[24] Bottom line: we need Congress to act.

SO, WHAT HAS CONGRESS DONE?

For more than thirty years, Congress has debated global warming—from 1992, when it ratified the United Nations Framework Convention on Climate Change, and every legislative session since. In 2009, the House of Representatives passed the American Clean Energy and Security Act to create a cap-and-trade carbon trading program, but it languished in the Senate. During the 117th Congress (2021–22), legislators introduced several climate bills, ranging from energy-efficiency programs, climate change education policies, a carbon tax, Build Back Better, and the highly publicized Green New Deal Resolution.

Ultimately, the 117th Congress passed aspects of the Build Back Better legislation with the Inflation Reduction Act of 2022 (P.L. 117–18). This law marks the boldest climate legislation and investment in clean energy in history. The Inflation Reduction Act creates more than $400 billion in tax credits and direct investment in clean energy, electric vehicles, environmental justice, consumer incentives such as tax credits and rebates for energy efficient appliances, rooftop solar, electric vehicles, and heat pumps. Many analysts claim that this legislation will put the nation on track to limit warming to 1.5 degrees Celsius by 2030.[25] We are making significant progress.

GREEN TOUCHPOINT

Think back to middle school civics class. Do you track federal and state legislation? Jot down what you would like to know more about or what you already know that you want to share with friends.

If you're a Spark or an Adventurer, your eyes might glaze over when you read a list of policy bullets. Remember, though, that legislation and market innovations can help yield the future we want. So here's a list and brief description of the primary federal and state policy strategies. For more detailed information, please visit www.onegreenthing.org/bookresources.

CAP-AND-TRADE	This policy would limit carbon pollution by creating a federally regulated market and declaring a price for carbon. Over time the government-determined "cap" on carbon becomes stricter and companies who violate their allotment would be penalized. The "trade" comes when companies who cut their carbon pollution can "bank" their excess allotment and sell it on the market, which incentivizes and rewards fast actors. Since the 1980s, the federal government has implemented a cap-and-trade system for air pollutants that cause acid rain, cutting these emissions by nearly 50 percent. China, Europe, and California have experienced carbon reductions through cap-and-trade.[26] Opponents to the policy argue that it's not fast enough to meet 2030 targets.
CARBON TAX	A federal carbon tax would be placed on the emissions companies produce. Taxes are simple to understand, easy to implement, and comprehensive. Furthermore, because companies already pay taxes, a federal system already exists to administer a carbon tax. Suggested prices for carbon range from $25 to $220 per metric ton, so the cost of the tax proposal and estimated revenue generated varies widely.[27] Critics say that a tax may not ensure emission cuts are reached as quickly as is needed and may incentivize businesses to move overseas. Some consider it "regressive tax" that would hurt underresourced communities who spend more money on energy-related goods.

GREEN NEW DEAL	Introduced as resolutions in the House and Senate, these broad principles support a major investment in green infrastructure, including clean energy jobs, health care, and job training. The goal is net-zero emissions by 2030. It calls for a ten-year mobilization to both push for 100 percent of our country's electricity to come from renewable power and upgrade the power grid. High-speed rail, electric vehicles, and retrofitting buildings are also key aspects. Critics argue that the resolution is too expensive and expansive.
REWIRING AMERICA PLAN	This plan declares that market-based solutions won't work fast enough. It envisions clear mandates, federal supports, and creative private financing to decarbonize the economy. As previously mentioned, the focus is on modernizing the electric grid to fund more solar and wind power as well as implementing electric vehicles, heat pumps, rooftop solar, and batteries. Opponents believe the plan is too costly and not a viable long-term solution.
FEDERAL CLEAN ENERGY STANDARD	The federal Clean Energy Standard would require utility companies to produce an increasing percentage of their power from clean energy, with a goal of 70 percent renewables by 2030. Proposed federal legislation would provide incentives for the development of renewable energy projects in BIPOC or underresourced communities and create a market for renewable energy credits to be banked, sold, or traded. Critics believe that market-based solutions, not legislative mandates, will create better outcomes.
THE INFLATION REDUCTION ACT OF 2022 (P.L. 117-18)	This legislation is the most sweeping climate legislation in history. It increased federal funding to more than $400 billion in direct investment and tax credits for clean energy technology, electric vehicles, grid modernization, renewable energy tax credits, and consumer incentives. Supporters believe this public investment will jump-start the clean energy future. Critics continue to argue that it's too expensive and expansive.

GREEN TOUCHPOINT

What proposed policy sounds the most promising to you?

These energy policies underscore an exciting aspect of our future. As Jamie Alexander of Project Drawdown Labs attests, "No matter where we work, every job is a climate job now."[28] That's why we must invest in climate literacy and climate change education. Several legislative proposals currently in Congress would also support integrated environmental education in K–12 and research universities.

COOL SOLUTION

Microgrids

A *microgrid* is a "relatively small, independently controlled power system that can be operated in concert with, or apart from, the local distribution and transmission system."[29] Microgrids can be powered by renewable energy and operate off the grid. College campuses and hospitals have used these types of systems for years. These hubs are becoming useful to metro areas and industrial facilities during emergencies. For example, during the 2019 California Camp Fire, the tribe of the Blue Lake Rancheria didn't suffer a blackout because of its resilient microgrid system. While two million people were left in the dark, this tribal government provided power for roughly 8 percent of Humboldt County. This energy security has resulted in the area becoming a cornerstone of the emergency fire response.[30] Ask about microgrids in your community as you consider zoning, development, and sustainability at the local level.

WHAT IS THE PARIS AGREEMENT?

In addition to US legislation, global efforts are being pursued, such as the Paris Agreement. The United States rejoined the Paris Agreement in January 2021. Ratified by 197 countries, it aims to reduce carbon pollution to net zero no later than 2050 by limiting the temperature rise to 1.5 degrees Celsius. Net zero requires countries to reduce carbon emissions and offset emissions through renewable energy projects, energy efficiency, or carbon sequestration, such as planting trees. The agreement also creates a Green Climate Fund to finance mitigation and adoption of carbon neutrality for developing countries.

CONSUMER OPTIONS: HOW YOU CAN MAKE A DIFFERENCE

Solar

Neighborhood solar gardens are a new concept in many states. The idea is that consumers "subscribe" to a solar installation owned by community members, including renters and those who can't afford solar installation. To learn more, visit the National Renewable Energy Lab website and its Guide to Community Solar.[31]

For homeowners, installing solar panels continues to become easier and more affordable. The average solar panel cost for a 3,000-square-foot home ranges from $15,000 to $30,000—but the prices for the panels and installation are decreasing, especially as competition increases. Businesses like Tesla, Sungevity, and Vivint now sell a variety of solar panel options. If the initial cost is a barrier, consider rent-to-own solar panels through groups like Mosaic, EnergySage, and SunRun, which provide financing arrangements.

In lower-income areas, the nonprofit Grid Alternatives is working to make clean solar energy affordable and accessible to ensure the clean energy revolution is grounded in equity.

Switch to Clean Energy

According to the Earth Day Network, 600 out of 3,300 utilities across the country allow consumers to switch energy providers.[32] Depending on where you live, you may have the option of switching to 100 percent renewable energy. This usually means your energy is generated in the same location, but your payment is used to buy renewable energy certificates to support green energy in other locations. The other option is to buy these certificates yourself, which ensures that for every kilowatt used, you purchase one kilowatt-hour of clean energy.[33]

Consider Buying Offsets

Consumers may directly buy carbon offsets for trips or commutes, and it's pretty easy. At the end of each year, I even purchase offsets for my family—which on average produces forty-eight metric tons of carbon dioxide equivalent—for around $280. Carbon offset companies sell credits or investments in projects that create carbon sinks. Think reforestation or soil conservation projects. Given that the average cross-country flight is about 360 kilograms of carbon per person, or $7 to $10 per cross-country trip, being mindful about travel can reduce your footprint.[34] Many airlines offer ways for consumers to offset travel when purchasing tickets. But beware—many offset companies have been accused of greenwashing and worse. Before you buy offsets, make sure the projects are verified by the EPA. Ensure that the type of projects align with your values by researching the details.[35] In my experience, buying offsets isn't very glamorous—no fanfare, big congratulations, or fancy certificate. You simply get a receipt. Make sure the company you're supporting reports back on how it's using your investment.

MOVE FROM HOPE TO ACTION

There are so many areas to learn about in the clean energy space. Transportation, electric cars, green building design, energy efficiency, and climate change education all have critical roles to play in the new clean, sustainable, and just economy. And I didn't even address hydropower, geothermal, the controversies around natural gas (and fracking!) and nuclear power, or international treaties on reducing methane and hydrofluorocarbons. My aim here was to hit some key concepts.

It's a lot to absorb, yet that's the beauty of the concept of One Green Thing: solving the climate crisis is a big task, but we can break it down into smaller steps. Your daily habits will drive the movement by changing our society's mindset and fostering a culture that embraces a clean energy economy. We must act as though our livelihoods depend on it, because they do.

Three Key Strategies to See Energy in a New Light

- Tell family members what you've learned about climate change and engage with your community about climate solutions.
- Research solutions that resonate with you.
- Lobby your local, state, and federal legislators for comprehensive climate change policies.

SEE ENERGY IN A NEW LIGHT ACTION CHECKLIST

Here are ideas for reflection and One Green Things, based on the Service Superpowers, that can help you move toward a clean energy future. For additional resources about seeing energy in a new light, visit www.onegreenthing.com/bookresources.

PROFILE	REFLECTION	ACTION
ADVENTURER	Pay special attention to how energy development impacts public lands.	Tell Congress that we need a clean energy policy now by calling your federal officials and writing snail mail letters, which tend to get more attention than email or phone calls.
BEACON	Strategize on how to organize around an exciting future vision.	Hold companies accountable by signing petitions, writing to the CEOs, and joining social media campaigns demanding action.
INFLUENCER	Consider how to explain global warming to a skeptic.	Call your utility company and ask about its renewable energy profile. Speak to a customer representative to find out how much of the energy that fuels your community comes from renewables.
PHILANTHROPIST	Focus on how to shape climate action through a service-inspired but practical lens.	Evaluate your financial portfolio to ensure your investments align with your values.

PROFILE	REFLECTION	ACTION
SAGE	Consider the barriers to access to energy for heating and cooling in low-income areas.	Urge your place of worship or community to switch to clean energy and engage with other faith-based groups on climate action.
SPARK	Take a close look at the consumer option section.	Attend your local school board meetings or email the board about the need for K–12 climate change education.
WONK	Ask yourself what climate solutions and policies excite you.	Email your city council or attend a meeting to ask about its sustainability plan and emergency preparedness protocol.

CHAPTER TAKEAWAYS

- Climate change is primarily caused by our energy and transportation choices from burning fossil fuels.
- There are multiple congressional proposals to address climate change, including the carbon tax, cap-and-trade, the Green New Deal, the federal Clean Energy Standard, climate change education, and many more.
- All the technology we need is available to reduce carbon pollution quickly. What's missing is the political and social will. That's why individual action matters.

JOURNAL PROMPTS

- Do you know where your energy comes from, or is it like magic to you? Have you ever talked with your energy company? Consider asking them about their portfolio and future clean energy plans. Write down three things you want to know about the energy in your community.

- Consider the unequal impacts of climate change. With record heat and traumatic weather events, the most vulnerable are likely in danger, either from heat stroke (without A/C) or freezing weather. What needs to change at the federal level? State level? In your community? Are you prepared for disruptive climate events?

- Do you have family members or friends who don't think the climate crisis is a problem? If yes, write down three things you'd like to discuss—compassionately and respectfully—with friends who aren't sure about the climate crisis. If no, write down three things you learned in this chapter.

Chapter 15

UNDERSTAND THAT YOU ARE WHAT YOU EAT

Rethink Food to Fuel Body, Mind, Spirit, and Planet

Whenever I'd visit my maternal grandfather, he'd make me an apple pie from scratch. My grandpa would pick apples from his small orchard and slice them up. He'd roll out the crust dough by hand, lay it in the pie pan, and sprinkle sugar and cinnamon on top. As a kid I didn't think much about this experience, but his apple pie was always delicious.

Then one day, at least twenty-five years later, I was apple picking with my family in an orchard in Winchester, Virginia. I was a super-busy mom, relishing time away from my high-pressure job to unwind in the country. We were laughing, filling our baskets with bright pink and crimson apples, and enjoying the crisp air that signaled autumn was near. As I checked into the old-timey orchard store to weigh our harvest, I smelled the apple pies baking. A flood of images came rushing back to me. I excused myself and found a bench to sit down, because I was overcome by loving, tender memories.

One particular revelation took my breath away. My grandpa had Parkinson's disease and a significant tremor. Most mornings it took him an hour to get dressed because buttoning his shirt was so difficult. Baking an apple pie must have taken him hours of intense concentration—and likely frustration when his hands wouldn't cooperate. Until that moment in the apple store, decades after he'd passed away, I'd never thought about how much gentleness, focus, and care had gone into that simple act. The pie I took for granted

GREEN TOUCHPOINT
Have you ever had a moment of breathtaking gratitude?

wasn't merely a dessert—it was a profound gesture of family, kindness, and service. Waves of extreme gratitude shook me.

Food is so much more than what we put in our bodies. It's culturally significant. It sustains life on this planet and is another aspect of how we're all interconnected. How your food is grown, where it's grown, and by whom it's grown are important layers to unpack and understand. Despite huge advances in food technology, global hunger remains an enormous threat, especially for children. Given that climate change will impact food production through heat, floods, droughts, plant diseases, and pests, it's time to rethink how we feed the world and care for the land.

Sparks will connect with the cultural aspects of food and appreciate greening school food. Adventurers, Philanthropists, Influencers, and Sages, you may be drawn to the exciting possibilities of regenerative agriculture and soil health. For Beacons and Wonks, reform of the system that governs food additives will likely pique your interest given the lack of regulatory oversight.

Let's start with what students are eating at school.

GREENING SCHOOL LUNCH AS A CLIMATE SOLUTION

When I was growing up in the 1980s, everyone wanted parachute pants, a Members Only jacket, or Guess jeans. Clothing was the visual indicator of someone's income. In contrast, my daughters could see the income disparity at their public school not by brand-name clothes or trendy sneakers but by the food their classmates ate at lunchtime. They packed their own lunches, which fit in a snazzy little stainless-steel container like a bento box. My daughter Cady would open her reasonably healthy lunch, then look around the table and see the ultraprocessed food that was served at school: frozen pizza, sausage on biscuits, and doughnuts. She once told me in a hushed tone, "Everything served at school is tan." It upset her because she knew some of her friends got most of their meals at school, yet the school food was gross.

As a family we discussed ways we could help. My husband and I talked about how in many communities access to healthy food is limited and explained the concept of "food deserts." Volunteering time and donating food (attention, Philanthropists!) can make a big difference but won't change the system.

We have to solve issues at scale, which is where school meals come into play. The national school lunch program costs $15 billion annually and serves meals to nearly 30 million kids a day during the school year. If we green school lunch in the United States, we can reduce carbon emissions to the equivalent of taking 150,000 cars off the road each year. That's why greening the supply chain of school food can transport us to a healthier, more climate-friendly future.[1]

How Did We Get Here?

The vast majority of kids in the United States consume most of their calories at school.[2] According to the National School Nutrition Association, before the pandemic, each day 100,000 schools served nearly 20 million free meals, with a total of 4.9 billion meals served annually. Hunger in America is a serious issue, and for many families, school meals were their only way to avoid hunger during the pandemic as schools shifted to drive-through meal pick-ups.[3]

The school lunch program is a critical lifeline for families and for children's health, but *what* food gets served has always been a hot-button issue. You may recall the 1980s controversy when efforts to cut billions from the school lunch program made it possible for ketchup to be considered a vegetable.[4] A change in the definition of "vegetable" allowed the US Department of Agriculture (USDA) to cut costs considerably because school districts could get rid of one of their two vegetable requirements. Because the cost of the lunch outweighs federal financial support, local school systems close the gap by outsourcing to private contractors. In the 1980s, only 4 percent of schools outsourced their kitchens, but by 2014 up to 25 percent did.[5]

Even after updating its nutrition guidelines in 2015 and offering more nutritious meals at school, the USDA gave in to industry pressure and allowed french fries and pizza. In April 2024, the USDA created new guidelines to reduce sugar in school lunch. While nutrition standards have improved, there's still much work ahead.[6]

How to Green School Lunch

How our food is grown, our personal health, and climate change are interconnected. Here are some great examples of ways schools are addressing these needs:

- **Greening school kitchens.** The nonprofit Eat REAL (think a green, sustainably grown certification for school meals) works with nutritionists to reduce sugar; increase whole-food options; create healthier, more plant-based and planet-friendly choices; and reduce food waste at scale. In 2020, Eat REAL helped the Mt. Diablo School District in California eliminate more than ten pounds of sugar per student and ensure that over one-third of produce served was sourced locally.[7]
- **Connecting farmers to schools.** The USDA Farm to School

program provides grants to introduce students to local farmers and understand where their food comes from.[8] FoodCorps enlists young people in service to work for school food administrators. Soul Fire Farm trains activists in Afro-indigenous farming and food justice advocacy and has sponsored more than one hundred community gardens.

- **Scratch cooking** in school kitchens, rather than using prepackaged food, is becoming more popular. For example, the Chef Ann Foundation provides a grant program to train food service systems in scratch cooking and reaches 75,000 students through twenty-one school districts.[9]
- **School garden programs** like Edible Schoolyards in New Orleans reported that participants have more interest in cooking at home after gardening. The Captain Planet Foundation's school gardens grew more than 100,000 pounds of fresh food for Atlanta families impacted by the pandemic in 2020.[10]

WHY YOU SHOULD KNOW HOW YOUR FOOD IS GROWN

Just as school programs are encouraging young people to learn more about where their food is sourced, it's important for all consumers to be more aware of this. How your food is grown impacts the environment, so agriculture is critical to addressing the climate crisis. The total agriculture sector is $1.1 trillion, constitutes 5.2 percent of the gross domestic product, and also contributes 10 percent of the US's greenhouse gas emissions.[11] After the Great Depression, the federal government

> **GREEN TOUCHPOINT**
> **What was the cafeteria food like in your school? What worked? What didn't?**

began supporting agriculture with "the farm bill." Passed every five years, this directs hundreds of billions of dollars to ensure fair pricing and protect the food supply. There is an extra incentive to produce corn via the ethanol mandate, which is governed by energy legislation.[12]

Although 75 percent of farm bill funding supports nutrition programs, each year our federal agriculture subsidy system gives billions of dollars in incentives to grow corn, soybeans, wheat, alfalfa, barley, oats, cotton, rice, peanuts, sugar, honey, wool, and mohair. These "commodity crops" mostly feed animals and are used to make processed foods. Only a small fraction of federal funds supports fruit and vegetable growers.[13]

The farmers I know, including my uncle, work hard to be effective stewards of the land. But large-scale, industrial agriculture requires tons of capital, large tracts of land, heavy equipment, pesticides, and lots of water. Conventional agriculture is hard on the soil and expensive. Unsustainable agricultural practices have released "billions of tons of carbon in the atmosphere," and have polluted water, air, and soil with toxic chemical pesticides, herbicides, fertilizers, and animal waste. Around the world, researchers are exploring how conservation practices can replenish carbon in the soil because "more carbon resides in soil than in the atmosphere and all plant life combined."[14]

The United States farm bill supports these soil conservation programs, such as the Conservation Reserve Program. Healthy soil practices include:

- Contouring, ploughing, and terrace farming to reduce water runoff.
- Planting perimeter trees and shrubs to prevent runoff and slow down water to infiltrate the soil.

- Planting cover crops, like radishes and turnips, rotated with other crops to suppress weeds and make sure the soil is covered all year-round and nourished with nitrogen.
- Letting the ground lay fallow instead of tilling it. This is an expensive technique, but repeated tilling can kill earthworms, harm microbes, and degrade soil.[15]

EQUITY IN AGRICULTURE

Historically, farm subsidies, low-interest loans, and crop insurance were wrongfully denied to many Black farmers. Extensive class action litigation and eventually legislation around the USDA's ongoing discrimination against Black and Native American farmers resulted in multibillion-dollar settlements.[16] But limited access to land, loans, farming equipment, and subsidies is still a barrier to Black farmers, whose population once represented 14 percent of farmers in 1920 and represent less than 2 percent today.[17] According to the Land Loss and Reparations project, Black farmers have lost $250–350 billion of "accumulated wealth and income" and "90 percent of the land they owned in the past 100 years."[18] You can learn more about the discrimination, as well as ongoing opportunities to support equity in agriculture, from the National Black Farmers Association.

Does Buying Organic Really Matter?

Yes, organic does matter. In the United States, the USDA Organic seal means that food is grown without hormones, toxic pesticides, synthetic fertilizers, radiation, or genetically modified organisms. Studies show that eating organic food can reduce the presence of pesticides in your body.[19] And because certified organic food is grown without toxic pesticides, it's better for the environment. Generally, foods with thick skins like avocados, onions, and squashes almost always show up as low pesticide residue options on the Environmental Working Group's (EWG's) Shopper's Guide to Pesticides in Produce, a great tool for choosing when to buy organic. Choose organic for leafy greens, apples, and strawberries. You can also consider shopping local or even trying out the 100-mile diet, only buying food grown within a 100-mile radius of where you live. (Refer to chapter 17 to see how toxic chemicals impact human health.)

Since 2008 organic sales have tripled, but land used in organic production hasn't increased as fast. Organic farming reflects only 0.6 percent of the production acreage in the United States.[20] Barriers to converting farms to organic are largely financial, which is why providing support to farmers to make the transition to either organic or regenerative agriculture can have significant climate benefits.

COOL SOLUTION

Regenerative Agriculture

Given that less than 1 percent of US agricultural land is used for organic production, there remains a tremendous opportunity to move toward regenerative agriculture. Sometimes called "carbon farming," it consists of practices like cover crops, no-till farming, using agroforestry with animal agriculture and crops, and adding beneficial microbes that replenish organic matter in the soil, enabling it to take more carbon out of the atmosphere and store more water.[21] The Rodale Institute predicts that regenerative agriculture could sequester more than the United States' total annual carbon dioxide emissions.

Who Grows Your Food?

Fifty-three-year-old Asuncion Valdivia died from heatstroke after working a ten-hour day picking grapes in 100-degree weather in the bright California sun. In 2021, federal legislators honored Mr. Valdivia's memory by introducing legislation to require national heat standard regulations to protect farmworkers from dangerous, climate change–fueled heat waves. More than two million workers labor in America's fields, picking crops or supporting dairy, chicken, and cattle operations. Approximately 73 percent of these workers are immigrants. Nearly half are undocumented. Exposure to toxic pesticides, crowded living conditions, and long hours means these workers are at high risk for injury. They don't receive overtime pay and rarely have access to paid leave. During the pandemic, farm workers had limited access to personal protective gear even though they were declared essential.[22] As we envision a greener, healthier future, we should work toward safer practices and better working conditions as well.

GREEN TOUCHPOINT

Have you spent time on a farm? What did you learn? If you haven't visited a farm, plan a visit soon to learn more about agriculture in your area.

What Are GMOs?

Genetically modified organisms (GMOs) are defined as "containing DNA that has been altered using genetic engineering."[23] Most of the GMOs in the United States are plants, including 90 percent of corn, sugar beets, and soybeans.[24]

GMOs remain controversial. One of the biggest issues is use of the toxic weed killer glyphosate with genetically modified plants, also known as RoundUp Ready seeds. These plants are designed to be resistant to the herbicide. Glyphosate kills the weeds nearby by blocking out an enzyme that regulates plant growth, but the genetically modified plant doesn't die once it's sprayed. Unfortunately, the weeds around the genetically modified plants have also become resistant to glyphosate. Therefore, farmers spray more toxic weed killer; there's been a 500 percent increase since it was introduced on the US market.[25]

The Interagency for Research on Cancer categorizes glyphosate as "a probable carcinogen" for humans. The EPA has declared that the chemical compound is *not* cancerous. On the other hand, the Agency for Toxic Substances and Disease Registry released a study showing exposure brought increased risk for non-Hodgkin's lymphoma and multiple myeloma.[26]

Glyphosate was presumed safe in humans because we don't produce the enzyme that the herbicide kills. However, some medical experts are concerned that glyphosate might contribute to "leaky gut" syndrome by killing beneficial bacteria and interfering with the protein zonulin, which affects the permeability of the intestine.[27] More research is needed for a definitive conclusion, but some physicians think that glyphosate exposure may cause Celiac disease–like symptoms and result in skin rashes, digestive issues, and allergies.

One thing's for certain: glyphosate is also showing up in our food—including hummus, milk, pasta, and cereal—and our bodies. In July 2021, after a $10 billion settlement, Bayer announced that it would no longer sell glyphosate by 2023, to quell the more than 125,000 pending lawsuits.[28]

You can reduce your exposure by changing your diet. *Environmental Health News* reported on a study demonstrating that only six days after switching to organic diets, 70 percent of participants experienced a reduction in glyphosate in their bodies.[29]

In 2016, after years of consumer campaigns, the United States joined more than sixty-four countries who require genetically modified food to be labeled. The USDA issued the rule in 2018 and by 2023, GMO products in grocery stores must be identified as "bioengineered," a relatively unfamiliar term for consumers, and through a QR code,[30] not an actual GMO label on the product. Some highly processed foods, like starches, oils, and sweeteners, are excluded. The law also provides a high threshold that allows 5 percent of ingredients to have GMOs. To avoid GMOs, choose organic foods whenever you can.

WHY EATING LESS MEAT IS BETTER FOR YOU AND THE PLANET

Trying a plant-based diet is one of the easiest ways to help the environment. Animal agriculture accounts for 20 percent of global carbon emissions. Project Drawdown calculates that if 50 percent of the world's population adopted a plant-rich diet, we could save 65 gigatons of carbon emissions globally.[31] Of course, a plant-rich diet is healthier too.

Campaigns like Meatless Mondays, where you skip meat at least once a week, are popular. Close to 70 percent of Americans are experimenting with more plant-based protein.[32] According to the Natural Resources Defense Council (NRDC), the ten most climate-damaging foods, in order from most to least harmful based on emissions required to produce them, are beef, lamb, butter, shellfish, cheese, asparagus, pork, veal, chicken, and turkey.[33] In addition to the environmental impact of animal agriculture, many are fed antibiotics to fatten them up and bring them to slaughter faster. This overuse can contribute to antibiotic resistance. The bottom line is to choose organic meat and support regenerative farms when possible.

Again, the goal isn't perfection—it's a daily practice of taking an action to make change. Start small and see what happens.

A NOTE ABOUT SEAFOOD

Globally, fish are experiencing threats due to unsustainable practices. Mercury in air pollution from coal-fired facilities pollutes lakes and streams, which means that many fish are high in this neurotoxin. Use EWG's calculator to make the best choices for low-mercury, highly sustainable fish. Safe bets include salmon, sardines, mussels, rainbow trout, and Atlantic mackerel.[34]

FOOD WASTE

My husband and I joke that Google documents keep the romance alive. We have a shared document and calendar for events and for . . . wait for it . . . the weekly menu. Menu planning has saved us so many times. Planning for reuse of meals or ingredients can reduce food waste and saves money and time. This simple trick has also immensely increased the quality of our life.

Food waste generates 8 percent of global carbon emissions. The *Washington Post* stated that "if food waste were a country, it would be the world's third-largest emitter of CO2, after China and the United States." More than 80 percent of food waste comes from consumers at home or from grocery stores and restaurants. While we can design better dining environments, provide smaller plates, address portion sizes, practice restaurant site composting, and repack partially damaged foods, the majority of food waste occurs at home.[35]

Community-wide and at-home composting can make an enormous difference. Even though I'm not a gardener by any stretch of the imagination, I do have a composter that gives me intense joy. I periodically shovel my compost on my backyard weeds. It makes me happy to see all my meals—while remembering that amazing online recipe I made or how I bought all those veggies for an abandoned juice cleanse—returning to the soil.

WHAT'S IN YOUR FOOD?

Now let's talk about one of my favorite things in the world: pie. My grandpa's pie was made of flour, water, baking soda, apples, sugar, and cinnamon. Contrast this to a frozen apple pie you buy at the grocery store. Filling: apples, sugar, water, modified corn starch, salt, spice, citric acid. Crust: enriched wheat flour (wheat flour, niacin, reduced iron, thiamine mononitrate, riboflavin, folic acid), soybean oil, hydrogenated soybean oil, water, dextrose, salt, dough conditioner (sodium metabisulfite). You need a dictionary to figure out how to pronounce some of these chemical names.

Chemicals in food additives have a long history in the commercialization of food. In the early 1900s milk manufacturers used chalk and plaster dust to make milk look whiter. Pesticides were added to canned vegetables to make them greener, and formaldehyde was used to preserve meat. As chronicled in *The Poison Squad,* USDA chemist Harvey Wiley pushed for food labeling after years of research and advocacy. Wiley enlisted twelve young male volunteers, whom the press deemed the "Poison Squad," to experiment with whether eating food with certain preservatives would damage their health. When the men invariably got sick, Wiley used the study to convince Congress to pass the Pure Food Act in 1906.[36] The point is that Congress took action only after scientific studies led to public outrage.

The Food and Drug Administration (FDA) regulates food additives. But out of the ten thousand chemical additives in food, only a small percentage have been publicly studied, and many companies don't have to disclose what's in their food products. Our food safety law allows a huge loophole, called "Generally Recognized As Safe," where companies can "self-certify" new chemicals as safe and put them in their products without telling the FDA.[37] About one thousand chemicals are in food without the company even notifying the FDA, which means we just have to trust the food companies to "make good choices."[38]

So how does this relate to climate change? Sugar, wheat, meat, and dairy—all energy-intensive products—form much of our ultraprocessed food products, which can contain largely unstudied additives and preservatives. Support groups like NRDC, EWG, and Center for Science in the Public Interest (CSPI) that advocate for food additive reform.

What's my advice? Eat real food, buy organic when you can, and read food labels.

HOW TO MOVE FROM HOPE TO ACTION

Food is more than nutrition. It can also be an artistic and cultural expression. It binds us to the earth and to each other. As you design your daily practice of sustainability, strive to make healthy choices for yourself and the planet. Even with a tight budget, you can create healthy and environmentally friendly meals. And who knows? Maybe twenty-five years from now, one of your family members will feel a tidal wave of gratitude as they remember you and think about a meal you prepared for them.

Some simple sustainable strategies include:

- Try Meatless Mondays.
- Choose organic when you can, especially for meat, berries, and leafy greens.
- Menu plan to reduce food waste.
- Compost or join a composting service.
- Skip low-sustainable, high-mercury fish.

UNDERSTAND THAT YOU ARE WHAT YOU EAT ACTION CHECKLIST

Here are ideas for reflection and One Green Things, based on the Service Superpowers, that can help you connect what's on your plate to protecting the planet. To learn more about how to understand that you are what you eat, visit www.onegreenthing.org/bookresources.

PROFILE	REFLECTION	ACTION
ADVENTURER	Flag the glyphosate discussion to connect the dots on weed killers, GMOs, and health.	Educate family and friends about soil conservation, regenerative agriculture, and why they're important climate solutions.
BEACON	Consider how to support sustainable agriculture and soil conservation.	Support children's health through advocating for universal school meals by writing, calling, or emailing members of your state legislature.
INFLUENCER	Focus on the regenerative agriculture section and its potential to create carbon sinks.	Call out Big Food companies on their failure to get rid of toxic additives. Share articles on the broken "Generally Recognized As Safe" system.
PHILANTHROPIST	Check out the organizations that support school lunch reform.	Spread the word about a farm-to-school or scratch-cooking program in your neighborhood.
SAGE	Contemplate how children's experience with food impacts their values later in life.	Encourage your congregation to adopt a plant-based lifestyle or plan a plant-based potluck at your next community gathering.
SPARK	Reflect on your food culture and how to support local growers.	Follow a nonprofit organization on social media dedicated to protecting farm workers.
WONK	Pay close attention to the organic section and the amount of land used in organic production.	Research the impacts of glyphosate on human health and the environment. Discuss with friends and ask the EPA to ban the use of this toxic weed killer.

CHAPTER TAKEAWAYS

- Food is more than nutrition. How your food is grown, where it comes from, and what you eat are significant factors in personal and planetary health.
- School lunch is a powerful tool to address climate change through the number of children fed, the need for good-quality food, and the market power to create demand for real food.
- The farm bill offers many options to ensure the conservation practices are sustained properly. Moving from conventional to sustainable agriculture can stop the cycle of overuse of pesticides, depletion of soil, and wasteful water practices.
- Get to know your local farmers and support environmental health and justice programs for migrant farm workers.
- Opt for organic food when you can to reduce the environmental impact and your exposures to genetically modified foods.
- Choose real food when you can and support efforts to reform the federal food additive system.

JOURNAL PROMPTS

- Reflect on your Service Superpower profile. What actions in "You Are What You Eat" resonate with you? What One Green Things are you already doing? What can you easily incorporate into your routine?
- Think about what you ate today. Where did it come from? How was it grown? How can you learn more or make better decisions for your health and the planet?
- What was a special meal or dish someone important to you made when you were growing up? What did it smell like? Look like? Taste like? What do you wonder about that meal now? Can you re-create it and share it with someone you love?

Chapter 16

PROTECT THE SOURCE

Renew Our Commitment to Protect Our Waters

I set my metal water bottle on the dais, tested the microphone, and sat back in my chair. The congressional hearing room was packed. When it was my turn to testify, my voice quavered for a second until I found my groove. I survived the requisite grilling from members of Congress on biofuels policy. But it's the stainless-steel water bottle that my colleagues remembered. The majority in the House of Representatives had recently declared that Styrofoam and single-use plastic were back in style in House cafeterias and hearing rooms.

My tiny act of rebellion—a One Green Thing—made a mark. My friends and I still joke about it today. You never know who will be inspired by the little things. One person's reusable metal water bottle won't solve the plastic pollution problem, global warming, or conserve the world's water resources. Compounded, though, the small ways we stand up for what's right can have far-reaching impacts.

WATER (AND YOUR WATERSHED) IS A LIFE SOURCE

How much do you think about the water in your life? Where does it come from? What's in it? If you live in a community like Flint, Michigan, you know how crucial clean water is. When I started studying environmental law, I was stunned by how much I took for granted. In the United States our drinking water is considered some of the safest in the world. Yet hundreds of communities in this country don't have access to clean water.

Our bodies are 60 percent water, and water covers 71 percent of the planet. Only 3 percent is freshwater, and 1 percent is drinkable. The simple definition of a "watershed" is "an area of land that channels rainfall and snowmelt to creeks, streams, and rivers, and eventually to outflow points such as reservoirs, bays, and the ocean."[1] Thinking about your community as a watershed is a compelling paradigm as you become mindful about the water in your life.

THREATS TO THE SOURCE

Global warming endangers our water supply. Extreme rainfall events create standing water that accelerates waterborne illnesses like cholera, typhoid, and diphtheria. Flooding jeopardizes access to clean water as waste ponds and water treatment centers become inundated and sewage overflows into lakes, streams, and rivers.

Cholera and raw sewage . . . this chapter is off to a rocking start! There's more where that came from—we'll talk about toxic algae, red tides, and dead zones too! Stay focused, team. Adventurers, pay special attention because water impacts all the natural spaces you love. And all of us

GREEN TOUCHPOINT

Similar to the chuckle my friend got from seeing my reusable water bottle in a congressional hearing room, have you ever been inspired by a small act of sustainability, kindness, or a habit of a friend?

can embrace the joy of learning new things, right? You'll soon see there are plenty of opportunities for you to get involved and protect the source.

Sewage Overflows

When I worked on Capitol Hill, the senator's office would get tons of calls when rapid-rain events caused water treatment facilities to overflow in Milwaukee, resulting in sewage overflows into Lake Michigan. If the untreated wastewater isn't released into natural bodies of water, it can cause sewage backups in people's homes. These sewage releases mean automatic beach closures. In August 2021, more than 380 million gallons of sewage overflow polluted Lake Michigan.

However, green design and infrastructure like greenways, riparian buffers, and rain gardens are dynamic tools to manage intense rainfall and stormwater runoff. In Milwaukee, for example, more than $4 billion worth of investments in rain gardens and green areas along the river have reduced overflows. According to the Milwaukee Metropolitan Sewage District, these overflows now occur only a few times a year, compared to forty or fifty times a year in the 1990s.[2] Across the nation outdated water utility infrastructure needs significant investment to protect public health, and this has the added benefit of creating more green spaces.

Toxic Algal Blooms

You know it's a fun read when we start talking about toxic algal blooms. Wonks, the next few pages is where you'll shine. Beacons, Influencers, Philanthropists, and Sparks—come along for the ride!

Warm water currents, low salinity, agricultural runoff, and fertilizers laden with phosphorus, nitrogen, and carbon all feed algae and generate overgrowth. When these tiny algae are out of control, they emit gases that contaminate the air and deplete oxygen in the water, making it toxic for wildlife and humans. As discussed in chapter 15, agricultural nitrate pollution demonstrates how the way we grow our food is connected to the planet's well-being and economic health. Communities who've experienced toxic algal outbreaks have spent more than $1 billion since 2010 on prevention and clean up.[3]

Consider the following:

- **Freshwater algal blooms** are common across the Midwest and associated with runoff from agricultural fertilizers. In 2014 in Toledo, Ohio, the Department of Health shut off drinking water for 400,000 people for three days because a blue-green algae called *cyanobacteria* contaminated Lake Erie.[4] In 2020, water for more than 500,000 Iowans was deemed virtually undrinkable because of an overgrowth of *microsytin,* a toxic blue-green algae "more poisonous than cyanide."[5]
- **Red tides** are caused by the bacteria *Karenia brevis,* a potent neurotoxin that turns ocean water brown and red and drops oxygen levels to dangerous levels for marine life. Nitrogen-rich fertilizers pollute ocean waters and are the biggest contributor to red tides. In 2018, a red tide off the coast of Tampa lasted for eighteen months, caused massive fish kills and the loss of $8 million from the fishing industry and tourism. Projected sea level rise of one foot around the coasts by 2100 means red tides are more likely, as coastal areas will experience warmer, more shallow water conducive to algal overgrowth.[6]
- **The dead zone** is an area slightly larger than the state of Connecticut (6,500 square miles) off the Gulf of Mexico that

has little or no oxygen in the water. Heavy rainfall and flooding cause nitrates to pour into the Mississippi River and the ocean. Dead zones also exist in the Chesapeake Bay and the Baltic Sea.[7]

EXTREME WEATHER: FLOODS, HURRICANES, AND DROUGHTS

Extreme weather events and flooding can result in contaminated drinking water, millions in property damages and loss of life, and a significant mental health toll. "Intensified short-term rainfall events are one of the hallmark manifestations of a human-warmed climate."[8] In Waverly, Tennessee, in August 2021, a record seventeen inches of rain fell in twenty-four hours. Twenty-one people died, including a twenty-three-month-old boy who was swept out of his mother's arms as she clung to a clothesline.[9] In July 2021, the city of Zhengzhou in China experienced "a year's worth of rainfall in 3 days." At least thirty-three people died, including twelve subway riders, and more than 100,000 people were evacuated.[10] Germany and Belgium also experienced historic flooding in 2021, resulting in more than two hundred deaths and an estimated $10 billion in damages.

Human-induced climate change is fueling more powerful hurricanes as well, because every degree Celsius increase in atmospheric temperature results in a 7 percent corresponding increase in the atmosphere's ability to hold moisture.[11] MIT atmospheric scientist Kerry Emanuel's research also demonstrates that "human-caused warming could lead to never-before-seen escalation of hurricanes, causing unheard-of surge in wind speeds of 100 miles per hour or more."[12]

The summer of 2021 also brought significant droughts throughout the West. Lake Mead experienced historic lows, leading to the first federal emergency water declaration that cut water allotments to Arizona, New Mexico, and Nevada. Water allocations were cut by 75 percent in California's Central Valley. In Northern California there wasn't enough water for farmers, tribes, and endangered fish in the Klamath Basin. Farmers in the Dakotas suffered as their crops dried up and cattle went thirsty.[13] These extreme weather events will continue to impact water quality and availability.

INDUSTRIAL POLLUTION: SLOW YOUR ROLL, ERIN BROCKOVICH

"Slow your roll, Erin Brockovich," exclaimed actor Maya Rudolph, who was playing a judge on the TV show *The Good Place*. After hearing that line, my daughter Cady's eyes widened. She suddenly became interested in my work with Erin on cancer clusters and started asking me questions. Funny how pop culture can make your kids look at you in a new light.

If you've seen the movie *Erin Brockovich*, you know that Erin's stellar investigative research, water sampling, and dogged determination resulted in one of the largest environmental contamination settlements in history. The polluter, Pacific Gas & Electric, settled for $333 million in 1996 after contaminating the water of Hinkley, California. The company used hexavalent chromium, classified as a "known human carcinogen," as a rust suppressor for natural gas pipelines, which then leaked into groundwater.[14] Exposure to this toxic chemical is connected to lung cancer, acute gastroenteritis, and liver and kidney failure. In her book *Superman's Not Coming: Our National Water Crisis and What We the People Can Do About*

It, Erin provides an outstanding manual on civic action in protecting your watershed and community. I encourage you to learn more through Erin's book: many American communities have suffered from industrial toxic water pollution, such as Camp Lejeune, North Carolina; Parkersburg, West Virginia; and Flint, Michigan.

Camp Lejeune

Camp Lejeune, known for training famous veterans who fought for our country, also holds another, darker legacy: it's home to the largest male breast cancer cluster ever identified.[15] More than 120 men diagnosed with breast cancer had one common thread—time spent working or living at Camp Lejeune and drinking its polluted water.[16]

The award-winning documentary *Semper-Fi: Always Faithful* chronicles the Camp Lejeune story through Jerry Ensminger, who lost his nine-year-old daughter to leukemia, and Mike Partain, who was diagnosed with breast cancer at age thirty-nine. More than a decade after his daughter died, Jerry heard a news broadcast explaining that families on base were exposed to toxic chemicals at levels 240 to 3,400 times higher than federal drinking water safety limits. Jerry "stopped dead in his tracks" when he heard the announcer say that the pollutants in Camp Lejeune's drinking water were associated with leukemia.[17] And Mike says a hug from his wife saved his life. She felt a lump and encouraged him to go to the doctor, where he received his shocking diagnosis. The legacy of water contamination at Camp Lejeune brought these environmental warriors together.[18]

On August 6, 2012, I had the honor of accompanying Jerry and Mike to the White House. In the Oval Office, President Obama signed a bill, the *Honoring America's Veterans and Caring for Camp Lejeune Families Act of 2012*, while Jerry, Mike, and the *Semper*

Fi filmmakers stood behind him. Because of Jerry and Mike's advocacy (Erin Brockovich also shined a bright light and lobbied for their cause), veterans who served on base from 1953 to 1987 and have diseases associated with the water contamination receive medical monitoring. Mike and Jerry are still hard at work. In March 2021, they petitioned the Department of Veteran Affairs to create a health registry to track residents of Camp Lejeune and address this toxic legacy. Certainly the fight for justice isn't over, but these men exemplify that persistence and individual action matter.

GREEN TOUCHPOINT

Is there a documentary or movie about a social justice issue that has changed your behavior? How did it make you think differently about that issue or about service?

Forever Chemicals and You

In Parkersburg, West Virginia, attorney Rob Bilott fought a twenty-year legal battle against chemical manufacturer DuPont. He uncovered corporate malfeasance and exposed a new class of dangerous, persistent chemicals, called "forever chemicals," that contaminate nearly every living thing on the planet.[19] (You can watch the exciting story in the movie *Dark Waters*, starring Mark Ruffalo.)

DuPont released toxic chemicals related to its nonstick line into the water, air, and nearby landfills. Company scientists monitored workers and found evidence of chemicals not breaking down and even passing through the placenta into the umbilical cord blood of newborns. DuPont researchers found correlations between exposure and the following health effects: cancer, thyroid dysfunction, pancreas and liver issues, colitis, and hormone disruption. DuPont quietly phased out the chemical but kept their essential health research secret.[20] In June 2023, Dupont entered into a $10.3 billion settlement with water utilities for polluting drinking water around the country with forever chemicals.[21]

The EPA uses the term "PFAs" (perfluorinated chemicals) to describe these forever chemicals, which include more than nine thousand substances. These chemicals are used in all kinds of products (because of their non-stick, stain-resistant, and heat-resistant properties) and are found in fire-fighting foam, pizza boxes, fast-food packaging, cosmetics, carpets, upholstery, rugs, clothing, and nonstick cookware. These nonstick chemicals are showing up in the air, water, soil, human and animal fat, and human blood. Studies show that 99 percent of Americans have forever chemicals in their bodies, and they likely pollute the drinking water of the majority of Americans, more than two hundred million people.[22] At least twenty-four states have taken action on forever chemicals as of January 2024, and pressure continues to mount for strong federal regulation.[23]

In March 2023, for the first time in twenty-five years, the EPA made the historic decision to recommend that a group of chemical contaminants be regulated under the Safe Drinking Water Act. You guessed it, they finally took action and chose these forever chemicals. Note that the EPA established a maximum contaminant level (the pollution limit) of 0.3 parts per trillion. The public health advisory that the EPA issued in 2023 set the standard at .004 parts per trillion–a *quadrillionth*—essentially declaring that there is no safe limit of forever chemical exposure in drinking water.[24] While EPA regulations continue to move forward and cleanup continues, the best protection of consumers right now is to fight for reform and take steps to reduce your exposure.

"Little Miss Flint" Protects the Source

When my girls reflected on young activists they admire, they told me about Little Miss Flint. Sixteen-year-old Mari Copeny, known as Little Miss Flint, launched her own water filter brand, raised nearly $500,000 in donations, and provided more than 16,000 backpacks with school supplies to Flint families.[25] She made national headlines in 2016 when President Obama accepted her invitation to tour Flint and signed a bill that provided $100 million to fix the town's outdated water system.[26]

The Flint tragedy has been called "the most egregious example of environmental injustice" because of the government's slow response, dismissal of residents' concerns, and the harmful lead exposure that Flint children suffered.[27] It started in April 2014 when the city decided to save money by temporarily switching its water source. Soon after the switch, residents began to complain about the odor and color of the water and health issues like rashes, headaches, and upset stomachs. They were ignored by local and state officials. After almost two years of exposure, the EPA issued an emergency order directing Michigan officials to take action because of the alarming levels of lead in Flint's water system.

Because of financial cuts, the city didn't properly treat the water to prevent water pipes from eroding. Therefore, lead leached from the pipes into the water and straight into people's faucets. The city's negligence exposed most Flint residents to dangerously high levels of lead. No amount of lead is safe, and exposure can result in damage to the nervous system, learning disabilities, and speech and hearing problems.[28]

Justice has come late to Flint with $400 million in state and federal aid and a $650 million settlement with Flint residents on behalf of the city. In January 2021, Michigan's attorney general announced

GREEN TOUCHPOINT

Think about Erin Brockovich, Rob Bilott, Jerry Ensminger, Mike Partain, and Mari Copeny. Is there an Eco Hero you'd like to meet or to recognize for their work?

that nine state officials, including former Governor Rick Snyder, had been criminally charged for their role in the Flint water crisis. The forty-two counts included perjury and involuntary manslaughter.[29] Though the charges were ultimately dismissed, historic and systemic racism, coupled with the drinking water trauma, underscore the ongoing lack of trust in state and local officials' assurances that the water is now safe.

COOL SOLUTION

Upland Forests in Cities

According to the Global Commission on Adaptation, restoring upland forests can help water utilities in the world's 534 largest cities regulate water flows,[30] manage more extreme floods, and save $890 million annually.[31] Green infrastructure serves as a carbon sink and manages overflows. This nature-based climate solution exemplifies how we can leverage green areas, sustainable design, and strategic investments to reduce carbon emissions, safeguard water quality, and protect human health.

HOW DID WE GET HERE?

The Safe Drinking Water Act regulates tap water in the United States by declaring legal limits for water contaminants. It establishes a standard to protect human health, called a "public health goal." Instead of meeting the public health goal, the law mandates a balancing of the economic costs of compliance to clean up the water pollutant. Therefore, just because it meets the legal limit doesn't mean it's safe.[32]

The EPA sets regulations for more than ninety different contaminants in public drinking water, including *E. coli*, *Salmonella*, and *Cryptosporidium*. Water utilities are required to test their water for these contaminants and disclose results to the public through Consumer Confidence reports. The rest are considered "unregulated contaminants," including forever chemicals and more than 160 other contaminants.[33] Some standards for regulated contaminants haven't been updated for fifty years. Scientific evidence has emerged that multiple unregulated contaminants in drinking water could cause harm at extremely low doses.

It's time for reform. The EPA should regulate water pollutants based on what's safe for public health and consider vulnerable populations like the elderly, pregnant women, and newborns. Water utilities need financial support to update infrastructure and water cleanup technology. The law should regulate whole classes of chemicals, not one chemical at a time.[34]

The Clean Water Act, on the other hand, regulates pollution at the source from factory pipes and from nonpoint sources like agricultural runoff. States and the federal government need more financial resources for water infrastructure and for oversight of nonpoint source pollution, which contributes to significant water quality issues.

IT'S RAINING PLASTIC

Plastic pollution is everywhere. According to the World Economic Forum, by 2050 there will be more plastic than fish in our oceans.

In addition to the deepest part of the ocean, plastic particles have been found in rainwater, Arctic ice cores, and snow in Antarctica.[35] Plastic also pollutes our bodies as we breathe in and consume plastic particles made of chemicals associated with plastic production, which we'll discuss in chapter 17. A study commissioned by the World Wildlife Fund found that we eat about a credit-card size worth of plastic each week.[36] Yuck.

Global production of plastic has increased dramatically as the fossil fuel industry deals with sagging markets. The increase of plastics production and associated incineration means that 10 to 13 percent of the global carbon emissions budget will likely come from plastic by 2050.[37]

Only 8 percent of the more than thirty million tons of plastic Americans throw away each year gets recycled. The rest is burned or goes in landfills. China is no longer accepting our twenty million tons a year of plastic waste, which means local landfills are drowning in plastics. Same with our oceans. The Five Gyres are gargantuan oceanic whirlpools that pull plastic fishing nets, buoys, consumer bottles, and plastic bags into large floating masses.[38] The largest of these gyres, called the Great Pacific Garbage Patch, is roughly the size of Texas and sits in the Pacific Ocean between Hawaii and California. Plastic is choking our planet.

Where Do We Get All This Plastic?

The majority of plastic manufacturing in the United States happens on the Texas and Louisiana coasts. Plastic production releases toxic chemicals in the air, like acetone, methylene chloride, styrene, benzene, and toluene.[39] The communities near these oil refineries that provide the essentials for plastic production are predominantly BIPOC. Called "fenceline communities," these towns are hurt by air and water pollution from local plastic production and experience higher rates of asthma, cancer, and environmentally related disease. That's why the plastics manufacturing zone along the Louisiana coast is known as Cancer Alley. Fighting plastic pollution is inherently linked to the fight for racial equality.

MOVE FROM HOPE TO ACTION

Climate change will create more pollution, extreme weather, and pressures on our water resources. Here are strategies to consider for a positive impact:

- To reduce nitrate pollution, which we discussed in chapter 15, we need to implement sustainable agriculture strategies that preserve soil, reduce nitrates in the water, and protect our oceans.
- To safeguard water quality and ensure flood control, we need to invest in green infrastructure in urban areas and update water infrastructure.
- To reduce plastic pollution, it's time to pass the national Break Free from Plastic Act, which would require companies that manufacture plastic to finance recycling and disposal programs and boost recycling and composting infrastructure.[40]

For personal actions to protect the source, consider the following:

- **Get to know your water utility.** Attend local meetings and learn about the specific issues in your community. Support funding for oversight and infrastructure upgrades.

- **Conserve water.** Take shorter showers. Buy water-efficient dishwashers and laundry machines. Use a water-efficient showerhead and toilet. Rethink your landscaping.
- **Know the top water contaminants.** Ten top water contaminants include lead, atrazine, PFAs, arsenic, GenX, hexavalent chromium, chloramines, fracking chemicals, TCE, and microplastics.[41] Learn more at EWG's tap water atlas or sign up for *The Brockovich Report*.
- **Skip bottled water and drink filtered tap instead.** Except for emergencies, bottled water isn't worth it. It costs two thousand times more than tap water. It isn't healthier either. A study found thirty-eight pollutants—including disinfection byproducts, radioactive chemicals, and industrial chemicals—in ten water bottle brands. The carbon footprint of bottled water is *3,500 times* worse than tap water. Check out *Consumer Reports* Guide to Water Filters to find what filter will work best for you.[42]
- **Reduce, recycle, and reuse.** Try to avoid single-use plastic, and use more real stuff.

PROTECT THE SOURCE ACTION CHECKLIST

Here are ideas for reflection and One Green Things, based on the Service Superpowers, that can help you connect with people across generations, so that together you can envision and move toward a greener future. For additional resources about how to protect the source, visit www.onegreenthing.org/bookresources.

PROFILE	REFLECTION	ACTION.
ADVENTURER	Reflect on how unsustainable agriculture and a lack of soil conservation can impact water quality and your recreation experiences.	Learn more about green infrastructure, stormwater runoff plans, and sustainable landscape projects in your town and how to support better water management in your city.
BEACON	Pay close attention to green infrastructure as a potential economic development tool and climate solution.	Next time you're outside, think about microplastics in the air, rain, and water. Share your knowledge with your friends and tell them about the Break Free from Plastic movement.
INFLUENCER	Research your local water utility and the water quality in your zip code.	Write to your favorite brand and ask them to reduce their use of plastic packaging in their products.
PHILANTHROPIST	Think about the essential role nonprofit advocacy plays in policy change.	Meet your local riverkeeper or watershed advocacy group. Volunteer for a river cleanup day or support environmental education on clean water in your community.
SAGE	Reflect on the challenges of the Flint community and mistrust of government.	Ask your congregation or community to dedicate a service to water conservation.
SPARK	Focus on the Camp Lejeune story and its community impacts.	Research PFAs and other toxic chemical contamination on nearby military bases. Share what you learn with friends.
WONK	Consider the limitations of the Safe Drinking Water Act and ways it can be improved.	Call your members of Congress to support stronger public health protections in the Safe Drinking Water Act and Clean Water Act, and provide support for water infrastructure funding.

CHAPTER TAKEAWAYS

- Water conservation is essential to the long-term health of people and the planet.
- Toxic algal blooms, red tides, and dead zones are created by agricultural runoff and will be made worse by warming temperatures. Soil conservation strategies can mitigate climate impacts.
- Human-induced climate change intensifies extreme weather events, including flooding, drought, and hurricanes.
- Toxic industrial pollution contaminates drinking water supplies across the country. We need to strengthen our national water policy.
- The Safe Drinking Water Act requires water pollution levels be set at the most economical technology for cleanup, not the public health goal. This means that legally safe water isn't necessarily safe.
- Plastic pollution contaminates our air, land, water, and bodies. Companies should be responsible for their products from design to disposal.

JOURNAL PROMPTS

- Is there a body of water—a lake, river, pond, or ocean—that is especially meaningful to you? What memories or feelings do you have when you think about it? What do you think the experience will be like in the future?
- Do you know about the water in your community? What watershed are you in? Where does your water come from? Take five minutes and Google it. Write down what you learned.
- Think about your Service Superpower and your strengths. How might you apply them to protect the source?

Chapter 17

VOTE WITH YOUR WALLET

Reevaluate the Sustainability of What We Buy

THE SCARIEST GHOUL EVER: PERILS OF A GREEN HALLOWEEN

In the third grade, Cady decided to be a ghoul for Halloween and wanted spooky zombie makeup for her costume. I suggested a non-toxic sunscreen as an alternative to cake foundation for her zombie look. After she and Susan trick-or-treated around the neighborhood, Cady refused to wear that chalky sunscreen ever again. When I'd try to apply it in the summer, she'd run away while exclaiming, "I look like a ghost!" Years later the girls heard my talk on toxic pollution in people, which haunted them. Sunscreen for Halloween makeup was as creative as this mom gets, but the girls were stunned once they learned why we used that instead of makeup from the dollar store.

Sustainable consumer products are becoming more popular, and understanding how toxic chemicals end up in consumer products and pollute the environment makes us smarter, better stewards of the planet.

POPS AND EDCS: INDUSTRIAL CHEMICAL POLLUTION CONTAMINATES NEWBORNS

How is that heading for a conversation starter? It's true and shocking. Yes, *industrial chemical pollution contaminates newborns.*[1] Fifty years after they were banned, chemicals like DDT (the pesticide made famous by Rachel Carson in her iconic book *Silent Spring*) and PCBs (a chemical that was banned in the 1970s) are still showing up in our bodies and the bodies of newborn babies. Those chemicals are persistent organic pollutants, also known as POPs, which bioaccumulate through the food chain and never break down.[2] Internationally, twelve POPs, known as the Dirty Dozen, are part of an international treaty because of their adverse effects on humans and the environment. These include pesticides, industrial chemicals, and chemical manufacturing byproducts.

In the past fifty years, childhood leukemia, brain cancer, and hormone-related cancers have exploded. A growing number of scientists, doctors, and pediatricians believe that this trend doesn't merely reflect better diagnostic methods, but that widespread, low-dose, chronic exposure to industrial chemical pollution contributes to these diseases.[3]

Many of these toxic industrial chemicals mimic estrogen and are called endocrine disruptors. These endocrine-disrupting chemicals, or EDCs, can act as hormones, increase or decrease the hormone production, or change one hormone into another. Like hormones, extremely low doses of EDCs can have more substantial effects on our bodies than larger doses.

The scientific community has known about the impacts of endocrine disruptors on human health for a long time. In the 1996 landmark book *Our Stolen Future*, the authors explained that these low-dose exposures to hormone-mimicking substances could impact obesity, fertility, intelligence, asthma, diabetes, endocrine-related cancers, and ADHD.[4] (Don't recommend this book to a pregnant friend. She won't be able to sleep for several days after finishing it. I made that mistake!) More than twenty-five years later, the peer-reviewed science demonstrates deleterious effects to our hormone system from these synthetic EDCs at lower and lower levels of exposure.[5] Common EDCs include the synthetic estrogen and plastics hardener BPA (bisphenol A), chemicals used in food packaging like PFAs, organophosphate pesticides, flame retardants, and drinking water contaminants like perchlorate and the herbicide atrazine.[6]

GREEN TOUCHPOINT

Are you familiar with POPs or endocrine disruptors? If not, are you surprised they're in consumer goods? If yes, what do you do to try to avoid exposure?

THE LINK TO CLIMATE CHANGE

Climate change will make toxic industrial pollution worse and expose us to more POPs and EDCs. Higher temperatures mean that toxic chemicals can vaporize and off-gas from polluted sites and plastics at a faster rate, or break down into toxic byproducts. Heat can also increase the redistribution of POPs that have settled in soils. Keeping POPs "locked in soil" is a similar strategy to creating carbon sinks. Given that many EDCs pollute drinking water, intense floods and hurricanes may increase exposure through additional sewage overflows and extreme rainfall events.[7] Firefighters spray flame retardants from airplanes to control wildfires, increasing the toxic load in water and soil.

HOW DID WE GET HERE?

In 2016, Congress passed a new law updating the broken federal toxics law, which was then widely viewed as the weakest environmental law on the books.[8] Companies lobbied for an updated regulatory scheme because consumer confidence was at an all-time low, and states had passed a confusing patchwork of chemical safety regulations. The new federal law mandates safety reviews for chemicals and requires that they be proven safe before going on the market. Now the EPA must protect vulnerable subpopulations, like children and pregnant women, in its chemical-review process. There are also some limits on companies' claims to keep information secret, so research scientists and academics can review chemistry and health studies with new chemicals. Unfortunately, states' abilities to regulate chemicals are insufficient, and the EPA has inadequate funding to enforce the bill.

Implementation of the new law has been dicey. It charged the EPA with reviewing ten priority chemicals, such as asbestos, dry-cleaning solvents, pigment violet 29 (used in paints and toy dye fabric), the degreaser carbon tetrachloride, and the flame retardant HBCD. To reduce the time for chemical review, the EPA "essentially rubber stamped" six hundred new chemicals and allowed them on the market with "little toxicity data." The EPA only evaluated the intended use, not the potential use, of the projects, which narrowed the scope of the review and failed to fully protect consumers. One thing is certain: consumers must remain vigilant about what they buy and bring into their homes and continue to demand strong chemical safety action from the EPA.[9]

CONSUMER PRESSURE MATTERS

A well-known example of how consumer action can affect manufacturing and policy is BPA. This synthetic estrogen was commonly used as an epoxy resin to line canned foods, a strengthener for polycarbonate plastic, and a coating for paper receipts. But BPA in cans leached into the food itself. More than two hundred studies linked BPA to hormone disruption, and research concluded exposure correlated to everything from cancer to reproductive problems, obesity, early puberty, and heart disease.[10] Bowing to intense consumer pressure and lobbying by environmental health advocacy groups, companies abandoned BPA. But some companies switched to problematic substitutes. So the products are technically BPA-free, but the substitutes can act as endocrine disruptors.[11] We need the EPA to review classes of chemicals, not just individual chemicals.

Phasing out flame retardants represented another win—at first. The three big culprits were PBDEs (brominated flame retardants), Firemaster 500 (containing a chemical similar in structure to a phthalate, which has been banned in children's products because of cancer concerns), and chlorinated tris (the chemical banned from treating kids' pajamas in the 1970s). A 1975 California fire safety rule resulted in widespread application of chemical treatments to prevent fires. Exposures to these "flame retardant chemicals have been linked to everything from hyperactivity/attention deficit disorder, thyroid disruption and even breast cancer."[12]

Scientific research, advocacy, and consumer action led to passage of a California law in 2014 to change the fire safety rule and promote safer consumer goods. This win shows we can make progress, but the problem of regrettable substitution still exists. In a 2020 study on new organophosphate flame retardants, scientists discovered that levels "are 10 to 100 times higher in air, dust, and water than the previous flame retardants." Study coauthor Dr. Arlene Blum of the Green Policy Institute stated, "It's disheartening that after years of health harm to our children from PBDE flame retardants, the most widely used replacements appear to be just as bad."[13]

These two examples demonstrate why policy reform, advocacy, and consumer pressure must dovetail. They're also a reminder that the work is never "done."

WHY THE TRUTH OF PERSONAL CARE PRODUCTS IS SO UGLY

Lead in Lipstick? Seriously? How is that possible? In 2011, the FDA tested four hundred different lipsticks and lip glosses and found lead in every sample. Although the average contamination was only 1.11 parts per million, there were 7.19 parts per million in a popular drugstore brand. The FDA said the lead levels weren't a safety concern, but the general consensus of public health experts says no level of lead exposure is safe.[14]

And that's just the beginning. A 2021 report showed endocrine-disrupting forever chemicals in a wide range of personal care products, from waterproof mascara to lipsticks.[15] Even asbestos is showing up in makeup through the talc in eyeshadow and blush. Asbestos is a known human carcinogen and is responsible for 12,000 to 15,000 deaths a year in the United States.[16] The FDA issued warnings in 2019 after asbestos was found in makeup targeted to teens. In 2020, women diagnosed with ovarian cancer successfully sued Johnson & Johnson for selling asbestos-tainted talc in baby powder.[17]

Switching to safer, less toxic products can reduce the presence of EDCs in your body. For example, a 2016 study examined the body burden of teen girls. When they switched to products with fewer toxic chemicals, there was a significant drop of these endocrine disruptors—specifically in their blood.[18]

Why do these studies matter? Some scientists think there's a connection between personal care products and early onset of puberty. Subtle hormone shifts in the body at levels as low as one part per billion or one part per trillion guide the complex onset of puberty. Research shows that early onset of puberty can lead to increased risk of depression, unwanted sexual attention, and the risk of dropping out of school.[19]

Key ingredients to avoid in personal care products include the following: parabens, phthalates, formaldehyde-releasing preservatives like DMDM hydantoin, skin lighteners, and diethanolamine (DEA). For sunscreens, choose titanium dioxide and zinc oxide, which provide strong protection and are the only two chemicals that the FDA declares as "generally recognized as safe and effective."[20] The term "reef-safe" isn't defined, so choose a sunscreen without oxybenzone (also an EDC) or octinoxate, which have been found to damage coral reefs.[21]

HOW DID WE GET HERE?

Why do consumers need to keep lists of ingredients on hand and scour small-print labels to make sure products are safe? Isn't that the government's job? And why are these toxic chemicals allowed in cosmetics in the first place? Isn't Washington all about overregulation? In some instances, yes, but with the personal care products industry, the answer is a decided no.

At the time of my writing this book, Congress last acted in any meaningful way on the cosmetics industry in 1938. That's right—almost a century ago. The only reason that cosmetics were included in the Food, Drug, and Cosmetic Act was because of the Lash Lure case. A wealthy Manhattan socialite went to get her eyelashes darkened and was blinded from a rat poison preservative used in the dye. A page and a half of law regulates the multibillion-dollar cosmetics industry.[22]

In the United States, the cosmetic industry has historically decided what's safe for us through the Cosmetics Industry Review Panel created in the 1970s to avoid regulation. Industry scientists advised the FDA on what ingredients should be restricted in cosmetics. This industry-led panel has recommended the restriction of a mere eleven ingredients in the United States. In contrast, the European Union has banned more than 1,300 ingredients from cosmetics.[23] The 1938 law assumed that cosmetic ingredients were safe before they went on the market. The FDA had virtually no authority to recall tainted cosmetics or force manufacturers to report cosmetic-related injuries; reporting is entirely voluntary.[24]

Cosmetics Reform

The good news is that in December 2022, the first major update to the federal cosmetic law since 1938 passed Congress. Called the "Modernization of Cosmetics Regulation Act of 2022," the new law provides FDA recall authority to get products that could hurt consumers off the shelf; requires labeling of allergens in fragrance; and requires companies to report adverse reactions to the FDA.

Recent legislative wins in California, the world's fifth largest economy, will result in far-reaching impacts in cosmetics safety. In 2021, two major bills passed the California legislature: the

Toxic-Free Cosmetics Act, which prohibits the use of toxic cosmetics ingredients, including parabens, formaldehyde, mercury, and forever chemicals, and the Safer Fragrance bill, which requires companies to disclose fragrance ingredients in cosmetics to the California Department of Public Health in a public database.[25] In 2022 and again 2023, the State of California amended the Toxic-Free Cosmetics Act to ban several classes of forever chemicals and several other potentially harmful ingredients, including "vinyl acetate, boron substances, and certain colors and styrene."[26]

New Brands Emerging

Although bipartisan support for cosmetics legislation reform exists, it's still extremely difficult to pass a new law in Congress. Perhaps you've noticed? Fortunately, companies are stepping up where the federal government has failed us. Brands like Tata Harper, W3LL People, and True Botanicals also offer clean beauty choices. Check out Black-owned clean beauty brands like the Laws of Nature Cosmetics, Luv & Co, and Rooted Women. But no matter the hype, check the labels. Unlike the "organic" label for food, there's no formal definition of "clean beauty" or "green beauty." The EWG Verified and Made Safe certification programs take much of the guesswork out of shopping. To stay updated on the latest news in cosmetics safety, you can follow organizations like Breast Cancer Prevention Partners, Safer Chemicals, Healthy Families, and the Campaign for Safe Cosmetics. Use the Skin Deep or Think Dirty apps to find better brands at the grocery store.

GREEN TOUCHPOINT

What are your favorite personal care products? Could you switch to greener, safer options? Or recommend a favorite green brand?

Why You Should Know What's Lurking Under Your Kitchen Sink

Study after study shows that chemicals used in the manufacture of household goods like stain removers, upholstery fabric treatments, and other cleaners end up in our bodies.[27] Here's an overview of some categories of household goods you should reexamine:

Flame retardants. Despite the phaseout of PBDEs, flame retardants are ubiquitous. Organophosphates have largely replaced them but as mentioned earlier, they share the same environmental concerns. Avoid polyurethane foam and consider using wool or mattresses with natural materials. Buy close-fitting children's PJs made from organic cotton.

Nonstick chemicals. These PFAs, more commonly known as "forever chemicals," are used in nonstick coatings, in food packaging, and to treat carpets and fabrics to make them stain resistant. This class of chemicals has been linked to obesity and heart disease. Avoid treated clothing, like rain jackets and all-weather wear. Consider checking your favorite outdoor brands on bluesign.com, which monitors companies' commitments to PFAs phaseouts.[28] Cook in a cast-iron skillet to avoid unnecessary exposure to PFAs.

Phthalates in clothing. These chemicals, found in cosmetics, plastics, and food packaging, are also used in clothing processes. Exposures to phthalates have been linked to hormone disruption, early onset of puberty, attention deficit disorder, diabetes, and lower IQs. Phthalates in fashion can also cause skin irritation, and children are especially susceptible.[29] Certifiers like bluesign and Cradle to Cradle are working with the fashion industry to make it more sustainable. These chemicals are hard to avoid, but choose natural fabrics when you can and wash your kids' new clothes before they wear them.

Household cleaners with toxic chemicals. Getting smart about the cleaners you use in your home can make it safer, particularly if you have young children. The Cleveland Clinic has a handy outline of key categories of cleaners to watch out for,[30] including:

- **Oven cleaners**, which usually contain toxic chemicals that may be fatal if ingested. These chemicals can trigger asthma attacks too.
- **Avoid fragrance** in general because it includes phthalates, which are hormone disruptors and can cause allergies.
- **Avoid "quaternary ammonium compounds,"** which are used to kill germs in fabric softeners and air fresheners. These ingredients can trigger asthma.
- **Laundry detergents** that have cationic, anionic, or non-ionic enzymes on the label can cause nausea or vomiting if ingested.
- **Avoid dryer sheets**, which are usually loaded with perfume and can contain the carcinogens benzene and acetaldehyde.[31]

COOL SOLUTION

Green Your Portfolio

If you have a retirement account or other investment accounts, talk to your financial advisor about your portfolio. Like many universities and other institutions, you can make the personal decision to divest from the fossil fuel industry. In addition to divesting, many financial institutions offer socially responsible or sustainable fund options to invest in green brands.

FAST FASHION

I have a confession. When I lived in DC, I wore fancy high heels with pointy toes and sported tailored suits. Living in Montana and surviving a pandemic has transformed my relationship to fashion. My current uniform is workout wear, now called "athleisure," which doubles as pajamas. Has the same thing happened to you?

Fashion plays an outsized role in global warming. According to the World Bank, 10 percent of global carbon emissions come from the fashion industry, "more than all international flights and maritime shipping combined."[32] If this sector's growth trends continue, the apparel industry will increase its carbon emissions 50 percent by 2030. From water pollution to questionable labor practices, fashion has received extensive global scrutiny. Nearly 90 percent of fiber used in fashion ends up in a landfill or is incinerated. The industry contributes one-fifth of the world's water pollution from fabric treatment and the dyeing process.[33]

Fast fashion revolves around ultracheap clothes usually worn only a few times, and it speeds up production time from factory to retail by producing fifty to one hundred microseasons a year. Over the past twenty years, people buy on average 60 percent more clothing.[34] Per the United Nations, a pair of blue jeans takes one kilogram of cotton, which needs about ten thousand liters of water to grow. That's the equivalent of drinking water for one person for ten years.[35]

Fashion brands and retailers are reducing carbon output by evaluating emissions from suppliers, investing in renewable energy and energy efficiency for their operations, and asking consumers to be more sustainable.[36] Fast fashion brands H&M and Zara have pledged to collect recycled clothing at their stores and increase their recycling efforts. Some companies, like LA-based Reformation and online consignment store thredUP, even report the carbon footprint of each product. Rental wear companies like Rent the Runway and Gwynnie Bee are encouraging a new model of a shared fashion experience. These examples demonstrate the circular economy, defined as "a model of production and consumption, which involves sharing, leasing, reusing, repairing, refurbishing, and recycling existing materials and products as long as possible."[37]

GREEN TOUCHPOINT

What simple changes could make your closet more sustainable or reflect the concept "less is more"?

We have an opportunity to change our behavior and send powerful market signals to the fashion world. Choose sustainable brands, buy used items, and repair your clothing. Consider thrifting or exchanging clothes with friends to reduce your carbon and environmental footprint.

ENVIRONMENTAL JUSTICE AND ACCESS TO GREEN PRODUCTS

As we discussed in chapters 15 and 16, most of the toxic petrochemical plants in the United States are near or next to Black, Indigenous, Asian or Latinx communities. In addition, few of these communities have access to greener, safer alternative products, which are sometimes not marketed to these consumers. As we embrace sustainability and a new circular economy, equity and justice are paramount.[38]

MOVE FROM HOPE TO ACTION

Engaging family and friends and helping them understand the broken toxics system might unsettle them at first. But small daily actions can reduce you and your family's exposure to toxic chemicals. Talk to your loved ones about asking how products are made and where they're sourced.

VOTE WITH YOUR WALLET ACTION CHECKLIST

Here are ideas for reflection and One Green Things, based on the Service Superpowers, that can help you vote with your wallet and support safer, greener brands. To learn more about how to vote with your wallet, visit www.onegreenthing.org/bookresources.

PROFILE	REFLECTION	ACTION
ADVENTURER	Consider alternatives to chemically treated outdoor gear and apparel to apply "leave no trace" in the context of a toxic legacy.	Check out the Green Science Policy Institute's PFAs-Free list for consumer products manufactured without forever chemicals.
BEACON	Reflect on BIPOC communities' limited access to less toxic consumer products.	Call your member of Congress and tell them to update the Food, Drug, and Cosmetic Act to ensure safe cosmetics.
INFLUENCER	Focus on consumer power in shifting brands and promoting a circular economy while we wait for policy change.	Identify your top five favorite sustainable brands, share them with friends, and ask for suggestions.
PHILANTHROPIST	Pay special attention to the nonprofit advocacy groups fighting for toxic chemical policy reform.	Donate and encourage others to support a fenceline community or environmental justice advocacy group.
SAGE	Note how consumer choices impact our physical health.	Ponder creation care and the impact of consumer goods on waste. Challenge your community to reduce its exposure to toxic chemicals and make a green switch.
SPARK	Strategize how to urge friends to switch to greener products to reduce body burden.	Walk friends or family through the connection between endocrine disruptors and climate change.
WONK	Enjoy the nerdy content on endocrine disruptors and climate change.	Explain the concept of a toxic chemical body burden to friends. Encourage them to choose less toxic products.

CHAPTER TAKEAWAYS

- Taking action on climate change means protecting our environment, which is inextricably linked to our personal health. Manufacturing products—especially consumer goods—uses a lot of energy and water and creates air, water, and plastic pollution.
- Despite recent reform efforts, toxic chemicals are still common in consumer products. These chemicals end up in our bodies and even small doses have been associated with a broad range of negative health effects.
- The Food, Drug, and Cosmetic Act hasn't been updated since 1938, and the cosmetics industry is mostly self-regulated. There are no clear definitions of "clean beauty" or "green beauty."
- Take action by reading labels and using trusted certifiers like Made Safe and EWG Verified, and follow the Campaign for Safe Cosmetics.
- Fast fashion is responsible for a considerable amount of global carbon emissions. Buy less, exchange, share, or purchase secondhand clothes.

JOURNAL PROMPTS

- What are some of your favorite sustainable brands? Consider sharing your recommendations with friends.
- What surprised you most about the content in this section? What do you want to learn more about? What will you share with friends?
- Think back to the 2030 visualization exercise in chapter 10. What will you be buying or not buying in the 2030 marketplace? Will sustainable or regenerative products be considered the norm? What habits do you have now that you think will be outdated by then?

Chapter 18

LOVE YOUR MOTHER (EARTH)

Revitalize Our Relationship to Nature and Wildlife

It was a beautiful August day in the Hayden Valley of Yellowstone National Park. Through my binoculars, I watched a bison herd cross the Yellowstone River, first one by one, and then about twenty bison crossed the deep water. The slanting sun baked the landscape in a golden light. Seeing these enormous creatures in the fast-moving river astounded me, not because they were elegant as they swam, but because I realized that I was witnessing a conservation miracle.

As it colonized the West, the US government purposely killed American bison to destroy Native American cultural, economic, and food systems.[1] When Europeans had first settled in the United States, 30 to 65 million American bison roamed North America.[2] By 1902, there were less than 200 wild bison left in the United States. That same year managers of Yellowstone National Park purchased twenty-one bison from private landowners and began a conservation restoration project. In 1907, the park managers established the now-famous Lamar Buffalo Ranch for bison recovery. Today Yellowstone National Park has roughly 4,500 bison within the park. Domestically, the US Fish and Wildlife Service estimates that more than 200,000 bison are in private herds.[3] In the early 1990s, the US government partnered with Native American tribes to transfer recovered bison to tribal lands. In 2016, Congress declared the bison the national mammal and proclaimed the first Saturday of November National Bison Day.

While recovery efforts are still controversial, this powerful conservation success story shows that conservation strategies work. We can make up for lost time. We have the science and technology. But we need the cultural and political will to protect our planet, wildlife, and ourselves.

Wildlife conservation protects human health and safeguards against climate change by teaching us compassion, protecting genetic diversity, and serving as carbon sinks as a nature-based solution to global warming. Additionally, connecting to nature is critical to our overall well-being.

HORSE SANCTUARIES AND CAT VIDEOS: WE NEED WILDLIFE, AND WILDLIFE NEEDS US

When my friend suggested we visit the United in Light Draft Horse Sanctuary in Livingston, Montana, I wasn't sure what to expect. The sanctuary provides refuge for sick, lame, or neglected draft horses. Once a month, you can visit the sanctuary to feed them, brush them, and feel their kind, gentle spirits.

As my younger daughter, Susan, and I headed out to the small field, we grabbed some brushes and found a raven-colored horse named Aribella, a thirteen-year-old Percheron mare. She was sixteen hands (more than five feet four inches high) and weighed almost two thousand pounds. This giant creature, who had experienced pain and neglect, now emanated serenity. As we brushed Aribella, we felt her relax, and so did we. Her quiet grace and enormous stature brought us into a mindful, peaceful moment of connection.

Animals remind us that we're not alone on this life journey. From household pets to therapy animals, research shows that animals make us more patient, more mindful, less anxious, and nicer.

Watching nature documentaries also reduces stress, alleviates anxiety symptoms, and leads to feelings of joy and wonder.[4]

A researcher at Indiana University's Media School published a study indicating that even watching cat videos on YouTube can increase happiness and decrease stress.[5] The study's author surveyed seven thousand different subjects who watch videos, asking about their personality, whether they owned pets, why they viewed cat videos, and how they felt after watching. The vast majority felt a sense of happiness from the videos. The author recommended that psychologists explore the concept of *digital* pet therapy.

GREEN TOUCHPOINT

Think back on an important animal experience, whether with a pet or wildlife. What do you most remember about it?

Stories about animal-human connections surprise me daily. A friend of mine attributes her discovery of cancer to her dog, who wouldn't leave her side and kept barking right in the area where her tumor was discovered. There are accounts of wildlife helping people escape drowning, scaring off intruders, and sniffing out Parkinson's disease and other ailments. And of course, there's the heartbreaking story from chapter 12 of this book of the boy and his faithful dog staying by his side until the end. We need animals, and they need us.

THE LOSS OF BIODIVERSITY, OR "THE SIXTH EXTINCTION"

I remember visiting the Field Museum in Chicago to hang out with my best friend. My daughter Susan was in third grade, and we were excited to see the dinosaur fossils—particularly the famous Tyrannosaurus Rex named Sue. As we were leaving the museum,

Susan became transfixed by a digital clock on the wall. I moved closer and saw that the clock recorded the number of species that had been lost since eight o'clock that morning. It was already at twenty-five. Tears welled up in Susan's eyes as she witnessed the Extinction Clock and realized that, on average, eighty-two species a day go extinct.[6]

The "Sixth Extinction" refers to the unsettling finding that we're likely in a phase similar to the other five mass extinctions the earth experienced over the past half-billion years. We can't necessarily stop it, but that doesn't mean we shouldn't try. Biodiversity is important because it protects human health, promotes planetary resilience, provides humans food and medicine, and fosters strong mental health. Around one million animal species and eight million plant species are at risk of going extinct "within decades."[7] Pollinators are at high risk, which is worrisome given that 75 percent of global food crops need bees and other insects to pollinate them. United Nations scientists have declared that the "critical window" for action to reduce biodiversity loss is the next ten to fifteen years, and that we've lost more than five hundred species in the past century, "a tally that would normally take 10,000 years to accrue."[8]

However, there's still hope. In the United States, the 1960s and 1970s brought about a complete environmental awakening. In 1962, Rachel Carson published *Silent Spring,* which documented the impacts of industrial and chemical agriculture on our waterways and wildlife. She alerted the public to the impacts of the pesticide DDT on transforming bald eagles' eggshells into brittle linings, thereby killing their babies by breaking during incubation or failing to hatch. The agricultural industry full-on attacked her, but she persisted.[9]

Rachel Carson's work created momentum for the EPA, which President Richard Nixon created in 1970, with strong bipartisan support for environmental action. Then Congress passed a slew of environmental laws from 1970 to 1980, including the Clean Air Act amendments, the phaseout of lead in gasoline, regulations on pesticides and insecticides, clean water protections, and toxic site cleanup and liability laws. I'm geeking out with this list because this decade demonstrated bold bipartisan leadership for the protection of our health and our planet. We worked together and made a lasting change for the American people, proving that it's possible.

The Endangered Species Act of 1973 fundamentally changed how our laws valued our relationship to wildlife and aims to prevent the extinction of listed species, which are declared as either "threatened" or "endangered." The law also protects "critical habitat" and prohibits what's called a "taking," which means to hunt, kill, harass, or otherwise jeopardize the species. We've seen big wins with bison, gray wolves, grizzly bears, red wolves, peregrine falcons, California condors, and bald eagles. Of the species listed, 99 percent remain on the planet today, and 68 percent are improving.[10] As these species recover many other issues arise, including creating wildlife corridors for them to move freely as they migrate. This issue is thorny in the American West, as grizzly bears begin to wander into subdivisions and wolves appear on cattle ranches. Even though these challenges exist, we have made conservation progress for wildlife.

GREEN TOUCHPOINT

Have you seen environmental progress in your life (e.g., cleaner air, water, or energy or a new park)?

WILDLIFE AND HUMAN HEALTH

Internationally, there is a substantial legal and illegal trade in wildlife. Millions of rare animals are smuggled across borders and sold. Scientists from the University of Oxford determined that from 2012 to 2016, 189 countries exported more than 11 million animals representing more than 1,300 species.[11] In the United States, which imports the most mammals and amphibians for pets, only a few animals crossing the borders are examined. As habitats are destroyed more animals crowd together in smaller spaces, which can result in "zoonosis" and make them more susceptible to disease. I bet you know where I'm going with this. Yep. The coronavirus.

Pandemic disease outbreaks, like coronavirus, and biodiversity are interlinked. According to the Centers for Disease Control, three out of four emerging viral diseases have spread from animals to humans. Called "zoonotic diseases," they transfer from animals to humans by direct contact or indirect contact like contaminated drinking water, food, air, or biting insects. For instance, the coronavirus is related to a similar virus found in bats. The Brookings Institution concluded that we must "fundamentally revise our relationship with nature" to prevent more and more pandemics on a global scale. The wildlife trade—both legal and illegal—results in transmission of infectious disease. Creating more space for wildlife, curbing illegal wildlife trafficking, and preserving habitat are essential strategies to reduce zoonotic diseases.[12]

Preserving genetic diversity also translates into greater potential for medicinal plants for human health and nutrition. The National Wildlife Federation underscores that "fifty-six percent of the 150 most popular prescribed drugs are linked to discoveries of natural compounds, with an annual economic value of $80 billion."

Think digitalis, which treats heart disease, the rosy periwinkle, which could help with cancer treatments, and the Pacific yew, whose compounds are already used in cancer treatments.[13]

Nature provides food for humans, and insects pollinate our plants. In addition to the plants we depend on for nutrition, hunting and fishing are essential skills that humans need to feed themselves, are culturally important, and are relevant as sport. I'm from Tennessee, my uncle used to be a taxidermist, and I live in Montana. Most of the hunters and anglers I know and have worked with are strong conservationists. Many have seen the impacts of climate change firsthand, with their favorite river or spot for their deer stand. They get it. Working with different constituencies lights the path to create lasting change.

Nature supports life as well as our economy. There's a wonderful term for this obvious statement: "ecosystem services." This means that clean air, water, soil, and land are essential to our economies and how human society operates. Plants also provide building materials like wood and rubber. Framing nature as "services" is an effort to break through the old "jobs versus protecting the environment" and shift to a "protecting the environment equals jobs" paradigm. By embracing long-term thinking, as we discussed in chapter 13, "Think Beyond Your Age," we can inspire others to adopt a new worldview that without a healthy environment our economies will fail.

NATURE AS A CLIMATE SOLUTION

Habitat conservation provides space for animals to roam, eat, and live, but it also provides carbon sinks with dense trees and protects against desertification. This concept is called "climate resilience."

The three most effective strategies are stopping deforestation, protecting soil, and limiting desertification.

Deforestation contributes 11 percent of all global greenhouse gas emissions caused by humans.[14] Demand for agriculture, mining, and housing leads to the burning of tropical forests, mostly in South America, to create room for grazing and development. The pressure on indigenous people in these areas is significant. When trees are cut down or burned down, the result is a triple whammy of carbon emissions because (1) the trees are no longer carbon sinks; (2) the trees release carbon when they're cleared by burning; and (3) the cut-down forests make room for livestock that create more carbon emissions.[15] Project Drawdown determined that curbing deforestation, especially in the tropics, could be a worthy climate solution.

Lands in dry areas of the world are turning into deserts as trees and soil are stripped away by unsustainable farming practices resulting in soil erosion, overgrazing, and faulty irrigation practices. Called "desertification," this process means that there is overall decline in land productivity, and "over 100 countries and more than 1 billion people in the world are facing the threat of desertification."[16] When grassland or forests are converted to agriculture, carbon stored in the soil is released, which increases carbon emissions. As we discussed in chapter 15, that's why healthy soils and soil conservation are so important to address climate change.

Carbon Offsets and Solutions

We can make a difference through nature-based solutions: planting trees, using sustainable agriculture methods, practicing soil conservation, and supporting coastal restoration. In addition to planting trees, we can promote soil health using the strategies discussed in chapter 15.

"Blue carbon" refers to ocean and coastal vegetation that stores carbon.[17] Think salt marshes, mangroves, and sea grasses, which store carbon in coastal soil. Coastal reforestation and wetlands mitigation techniques can be powerful tools in climate solutions. In wetland areas, mangroves are particularly important as conservation strategies.

You may have seen ads for carbon offset projects. Both companies and consumers are buying carbon offsets. Carbon offset projects are frequently tree planting, but sometimes they're soil conservation systems, including eliminating till agriculture, growing kelp, or planting mangroves. Other projects include biogas digesters, organic waste management, solar water management and compost, and wind and solar power. Stricter regulations for the carbon offset industry have been proposed to protect consumers from greenwashing.[18]

TIME IN NATURE PROMOTES OVERALL WELL-BEING

I heard the phrase "go outside and play" every day of the summer when I was growing up, and I bet you did too. Study after study shows that time outside makes us happier and less stressed out. Even five minutes outside can lower blood pressure and increase our sense of well-being. Kids are constantly connected to digital devices—some even up to ten hours a day. Being online and indoors 24-7 is bad for our collective mental, physical, and spiritual health. Humans are hardwired to connect with the outdoors. Simply viewing photos of nature can still the mind and help us relax. Key benefits of spending time outside include improved physical fitness, more vitamin D, enhanced sleep, decreased anxiety, lowered cortisol levels, reduced ADHD symptoms, and increased creativity and self-esteem.[19]

Time outdoors provides opportunities for us to enjoy each other and the landscapes around us. You don't have to be an Adventurer or own a lot of gear or engage in a specific activity to play outside. You also don't have to plan a lengthy, expensive vacation. Loosen up, make time for yourself, and simply try to connect with nature each day. Yes, it's easy for someone like me, who lives in Montana, to go on and on about spending time in nature. Rest assured that you can connect with nature no matter where you live. The mere act of taking a walk outside has been called a "superpower" by neuroscience researchers who have shown this activity reduces stress, increases fitness, and lifts our spirits.[20]

In addition to downtime in nature, environmental education helps students achieve in the sciences and results in higher overall test scores and better critical thinking skills.[21] During the pandemic, schools experimented with outdoor classrooms and have seen significantly more engagement from students. More school systems across the country are investing in environmental and outdoor education, like in Oregon, Maryland, and California. Hunting and angling classes for kids and bringing back recess and outdoor play as part of school are also key strategies to connect children to nature.[22]

EQUITY AND NATURE

Many of the public lands that environmentalists like me hold so dear are ancestral lands of Native Americans. Recognizing this legacy means that we should tell how our parks were created. For example, twenty-six different tribes consider Yellowstone sacred communal ground. Acknowledging our diverse history and the ways we view nature, wilderness, and public lands shows the dynamic experience of connecting to nature and its relevance in our culture. We should also recognize that for many people, nature and public lands aren't viewed as safe or inclusive places. That's why we need to increase access, provide environmental education opportunities, and talk about our shared history.

OUR PUBLIC LANDS & SOLASTALGIA

Managed by the US Department of Interior, our national park system extends to all fifty states, the District of Columbia, and US territories, with 423 individual units consisting of more than 85 million total acres. The Bureau of Land Management (BLM) and the Agriculture Department's US Forest Service also manage millions of acres of inspiring landscapes on our behalf. The Congressional Research Service estimates that the federal government owns around 28 percent of the United States (640 million acres).[23]

And of course, global warming will change these landscapes significantly.

Take Yellowstone, for example. In 1872, President Grant signed into law legislation creating Yellowstone, the world's first national park, for the "benefit and enjoyment of the people." This magical area has ten thousand thermal features, more geysers than anywhere else in the world. It also contains Yellowstone Lake, the Lamar Valley, and Hayden Valley—some of the most spectacular scenery in the United States. Called America's Serengeti, it's known for its remarkable wildlife, as well as herds of bison, wolves, grizzly bears, elk, pronghorn, and sandhill cranes.

GREEN TOUCHPOINT

What's your favorite national park or public land? What do you know about its history? Do you know what tribes are associated with that land?

Like many national parks, this famous ecosystem is under threat from climate change.[24] Since 1948, Yellowstone temperatures have risen nearly two degrees. This warming means big changes are in store, including likely changing the plant composition in Yellowstone. The growing season has expanded by more than thirty days, which will likely lead to more invasive plants like cheatgrass. The range of keystone species like the white bark pine, which serve as food sources for the Clark's nutcracker and grizzly bears, will become smaller. Scientists report changes in bison migration and fish spawning as waters rise and spring begins earlier. The shorter winter will likely extend wildfire season. There is little doubt that Yellowstone will still be here in another 150 years. But the question remains: What will it be like for our great-great-grandkids and the wildlife? Philosopher Glenn Albrecht calls this feeling of longing or homesickness "solastalgia."

COOL SOLUTION

Blue Carbon Wetland Protection

Wetlands and mangroves serve as buffers and natural filters for water contaminants and also soak up carbon. The recent MarVivo Blue Carbon Conservation Project in Magdalena Bay in Baja, California, will be one of the largest blue carbon projects in the world. It will manage 22,000 hectares of mangroves and is projected to reduce carbon emissions by 26 million tons of carbon dioxide equivalent over thirty years.[25] The $6 million project expects to recoup $2 million in direct annual benefits to local communities.

A NEW ETHIC? THE RECIPROCITY PRINCIPLE

Our Wild Calling by Richard Louv brings forth a powerful concept of environmental stewardship that all the Service Superpowers, but especially Sages, will appreciate. As we think of what life will be like in the future, it's worth examining Louv's Reciprocity Principle:

> For every moment of healing that humans receive from another creature, humans will provide an equal moment of healing for that animal and its kin. For every acre of wild habitat we take, we will preserve or create at least another acre for wildness. For every dollar we spend on classroom technology, we will spend at least another dollar creating chances for children to connect deeply with another animal, plant, or person. For every day of loneliness we endure, we'll spend a day in communion with the life around us until the loneliness passes away.[26]

The Reciprocity Principle parallels chapter 12 of this book, "Know You Can't Go It Alone." Connecting to others and the life around us can help us become better, kinder, more wholehearted people. Consider comparing the principle to your One Green Thing "why" from chapter 2 and seeing what elements you might want to incorporate.

HOW TO MOVE FROM HOPE TO ACTION

To love our mother (Earth), we need to take time out and focus on the ways the planet sustains us, our connection to all living things and each other, and the possibility for a brighter future. The changes we need must come fast. But as the Laws of Change underscore, the journey starts with daily actions to create the momentum for lasting impact and a seismic cultural shift.

LOVE YOUR MOTHER ACTION CHECKLIST

Here are ideas for reflection and One Green Things, based on the Service Superpowers, that can help you love your mother (Earth). For additional resources about how to love your mother (Earth), visit www.onegreenthing.org/bookresources.

CHAPTER TAKEAWAYS

- Despite the grim news about rapid species extinction, we have positive conservation success stories to provide hope and demonstrate that change is possible.
- Time with animals makes us happier, more compassionate people. Research shows that even watching nature videos can increase compassion.
- Endangered species trade is linked to global pandemic and zoonotic disease. Therefore, protecting wildlife also safeguards public health.
- Climate change threatens wildlife habitats around the world. Rapid deforestation contributes to global warming.
- Nature-based solutions like tree planting and blue carbon can conserve habitat, reduce carbon, and support local economies.

PROFILE	REFLECTION	ACTION
ADVENTURER	Contemplate public lands, westward expansion, and Native American history.	Research Native American connections to the public lands you value. Share what you've learned with friends and family.
BEACON	Identify your favorite nature-based solution.	Encourage friends and family to support local land trusts to foster conservation in your community.
INFLUENCER	Reflect on eco-anxiety and the role of outdoor education.	Share what you've learned about eco-anxiety on social media. Encourage friends to discuss the connections between mental health, access to nature, and the climate crisis.
PHILANTHROPIST	Contemplate how natural areas in your community support the local economy.	Participate in or plan a day of volunteering to support a local park or waterway.
SAGE	Consider the spiritual implications of Louv's Reciprocity Principle as you review the research on mental health and public spaces.	Support a local animal welfare organization and encourage your community to work to protect wildlife and wild places.
SPARK	Look for positive conservation stories to counter fears of rapid species extinction.	Talk to friends about the connection between habitat loss, climate change, and climate resilience.
WONK 	Think through how public policy might support wildlife in a changing climate.	Contact your local Fish & Wildlife agency or visit its website to learn more about climate adaptation policies in your community.

JOURNAL PROMPTS

- Think about what animals mean to you. Did you have a special pet or connection with an animal that helped bring you comfort, shaped who you are, or enabled you to enjoy the awe of nature? If not, have you experienced a moment, like mine in Yellowstone or at the horse sanctuary, that connected you in a meaningful way to the natural world?

- Did you have a favorite outdoor spot you'd visit when you were a child? Describe what it was like. What made it special? How did you feel in that space? What did you see, hear, smell, or taste?

- Reflect on Louv's Principle of Reciprocity. Do you have an ethic or principle that reflects your conservation ethic? Revisit your One Green Thing *why* from chapter 2. Does it still resonate after reading these chapters on how to effect change? If yes, why? If not, consider revising it.

Part 3

HOW IT ENDS

Chapter 19

IT'S UP TO YOU

In my late thirties I bought a Wii Fit, which was all the rage. Instead of virtual bowling and baseball with my kids, I was ready to embark on my own life-changing fitness journey with merengue, salsa, and aerobics. I'd step on the console, and it would blurt in a loud, obnoxious, electronic voice: "Heather, your weight is ______. Your real age is thirty-nine, and your *medical age* is fifty-four." To make matters worse, the game would provide verbal feedback when I missed a move with a shrill refrain telling me that I had failed. For the life of me, I couldn't figure out how to mute the dang thing. The gaming console was situated in the living room, which meant my family was thoroughly entertained by my annoying experience. So like any mature, recovering perfectionist, I banned everyone to their rooms while I was exercising. After a few workouts, I gave up and never used it again. I'm all for personal growth, but there's not much fun in hearing how much you suck.

Let this book be the complete opposite of what I experienced with that awful Wii Fit. Some people expect a refrain of shame when they miss a sustainability beat or assume that any missteps will be broadcast loudly and with disdain. That's not what *One Green Thing* is about.

Your superpowers, creativity, and unique talents are valued. Being part of this community can add more fun, compassion, and beauty to your daily life and create a brighter, more sustainable future. The compounding action of your sustainability habits and engaging your Service Superpower matters. Stay with it, because we're in an all-hands-on-deck moment for the planet. We still need to laugh. We still need joy. We still need hope. We can take the issue seriously without taking *ourselves* too seriously.

My daughters have modeled putting their Service Superpowers in action. Two years after I offered to *drive* my older daughter, Cady, to her first climate strike (as I recount in chapter 13), Cady organized the Bozeman High climate walkout. As a Beacon, she recruited key leaders in the school to protest. I watched in admiration as she and my younger daughter, Susan (a Spark), walked from the high school to the Gallatin County Courthouse steps. It was a cloudless fall day. Golden yellow leaves swirled around the teenagers as they held up their handmade posters demanding change and chanted: "We are unstoppable! Another world is possible!"

Cady started the rally with a land acknowledgment to recognize the tribal land that the courthouse sits on. Then her speech made my jaw drop with its passion and eloquence. As the *Bozeman Daily Chronicle* reported, she also stated: "I wish I could worry about my homecoming outfit or who asked who out or even just my calculus homework. But instead, my main worries are how to save humanity from itself—how I'm supposed to solve the impending apocalypse that people don't even believe is happening."[1]

As a mom I was proud that my daughters took their anxiety and channeled it into action; both girls felt energized by the walkout. I also know that if we don't act globally, Gen Z will experience three times as many climate-change disasters as their grandparents, and seven times as many heat waves.[2] The majority of the carbon pollution in today's atmosphere was created after 1990. I graduated from high school in 1991. That means that greenhouse gas emissions produced in my lifetime will impact my kids and future generations more than they ever hurt me.[3] Those who have contributed the least carbon pollution will suffer the most. Although global warming is scary, the collective change we can make and the future we can build is exhilarating, exciting, and wondrous. As these kids chanted at the strike, another world *is* possible.

In the foreword, Erin Brockovich reminded us that we are the superheroes the earth needs right now. By showing up in service, embracing compassion, and speaking truth to power, we can become the awesome ancestors that future generations deserve.

Whether you are an Adventurer, Beacon, Influencer, Philanthropist, Sage, Spark, or Wonk, apply the tools provided in this book. Complete your 21-Day Kickstarter Plan, then design your own Eco-Action Plan to continue your new habits and track your joy as you engage in your daily practice of sustainability. Join the OneGreenThing community, focus on the seven areas to effect change, and know that your Service Superpower can help us create a healthier, greener, more equitable future.

When I was finishing this book, one night at dinner I asked the girls if they had any advice for Millennials, Gen Xers like me and my husband, or Boomers like their grandparents.

Cady sighed. "I've said it all, Mom."

But Susan quickly replied, "My advice is simple: *listen to the kids*."

As we move through life, we transform into different generational roles. I remember writing to my parents from Lake Tekapo in 1994 about my passion and decision to become an environmental lawyer. Now I'm middle-aged and will soon have college-age children. My parents are grandparents now. It's becoming clearer to me that "we're all just passing through this beautiful earth," as my grandfather—the one who made me apple pies—used to say.

As you reflect on your personal journey and as you pass through this world, what will your legacy be? How will you apply your Service Superpower? What will your future loved ones thank you for?

You can start today.

Right now.

We need you.

And so do they.

You write the final chapter of the future.

ACKNOWLEDGMENTS

Principle number one of One Green Thing is "Know You Can't Go It Alone." To the reader who picked up this book, gave it to a friend, or checked it out of the library, thank you. We need you in the climate movement.

Thank you to my wonderful husband, David. You are my everything. My daughters, Cady aka "Pal" and Susan aka "Ace," thank you for allowing me to share your stories and for teaching me how to listen.

Amy Hughes, thank you for your determination and belief that we could create an accessible, fun book to address the existential threat to life on the planet. You are a thought partner, counselor, and dear friend. Andrea Fleck-Nesbit, thank you for this incredible opportunity to share my story and Matt Baugher for your dedication to this project. Amanda Bauch, Meaghan Porter, and Austin Ross, thank your for your intelligent edits and tireless effort. To Kevin Smith, Lauren Kingsley, Hannah Harless, Sicily Axton, thank you for the expert marketing help. To Molly Stratton, graphic designer extraordinaire—thank you for making the Service Superpowers come to life. Emma Loewe, thank you for introducing me to Amy.

Erin Brockovich, you've been an amazing friend for more than a decade. The "Brockovich hypothesis" keeps me laughing and enjoying life during tough times. To my team—Emily Fowler (and the whole family!), thanks for your countless hours of editorial advice, counsel, and detailed feedback, and all your encouragement to make this book and OneGreenThing, the nonprofit, a reality.

To my parents, Sue and Jim, thank you for taking me outside when I was young, so I could understand my duty to take care of God's creation. Throughout my life, your advice has always been aligned, and I'm grateful to you for that and your unconditional love. To Mike, Penny, Kathleen, and Barry—thank you for being there every step of the way. To the late Nick Marler, I'm grateful for all our adventures. Thank you to my extended family (Eric, Sherry, Mark, AJ, Gail, Donnie, Christine, Liz, Dan, Jane, Greg, Pat, Clarence, Boddie)—and especially Aunt Vickie, who took to me to Jamaica to get this book going—I love you. To the Brenner-Diamond-Mitz Shakespeare Zoom group, big hugs. To my nieces and nephews, especially the late Mantha Rabon, thanks for all the memories and intergenerational fun.

Thanks to the OneGreenThing board of directors: Jen Boulden, Erin Brockovich, Leslie Carr, Gigi Lee Chang, Julia Cohen, Will Cole, Chris Colin, Meg Dees, Max Finberg, Victoria Gilchrist, Kelly Herman, Chris Hill, Robin Hill-Emmons, Jen Fisher, Asher Jay, Karin Johnson, René Jones, Nora La Torre, Robyn O'Brien, Maya Penn, Larry Schweiger, Sara Sciammacco, Mindy Simon, Pam Solo, and Colleen Wachob. And thanks to the OneGreenThing advisory council: Dr. John Wear, Dr. Emily Elliott, Eleanor Bigelow, Leslie Everett, Ali Fowler, Dr. Myleen Leary, Paul Hashemi, Karla Raettig, Adam Olenn, Nisha & Vijay Raghavan, Skye Raiser, and Nancy Stabell.

To the Catawba community—John, Cathy, Marty, Meg & Jay, and Fred & Alice—thanks for cheering me on. To Elaine, James, Kelly, Leslie, Meredith, Megan, Nils, Thomas, and Ty, thank you for your professionalism, tremendous talent, and friendship. To my talented former professional colleagues in the environmental community and on Capitol Hill—keep up the stellar work!

To all the Eco Heroes profiled in the book—thank you for sharing your inspiring stories. To Mary Frances Repko, you will always be an Eco Hero to me! To my teachers, especially Marcia Freas Lysher and Beverly Fairchild—thank you for all you taught me and for your lives of service. To Naomi—you were right.

Nisha, Susan F., and Susan D.—the Kilimanjaro story is for you. Sally, Lauren, Amy, Chris, Paul, Kyle, and Hunter—thank you for the New Zealand adventures. To my circle of love in Takoma Park (Laura, Ingrid, Kate, Thembi, Mary)—miss you. To Amy S.—the breakfasts at Busboys & Poets were worth it. To the Hot Moms of Bozeman (Jamie, Patty, Valerie, and Susy)—your friendship, wit, and kindness got me through the pandemic. To Myleen and Asher, thanks for the "hive" sessions. Mindy and Nancy, our Park City trip was pivotal. Susan G., thank you for being there through all of life's phases from high school to midlife! To Victoria, whom I met on the first day of college, thank you for all your prayers! To 2300-A Price Ave. (Isabel, Cooper, and Emily)—thank you for reminding me that life is beautiful, messy, hard, and magical all at the same time. To Paul, Mary, Cate, and Alex, thank you for all the summer memories. To Jae—I've loved being your friend from eighth grade English to the Beinecke and beyond. To my favorite doctors, Reena and Jen (and Deepak!), thank you for always being there for us. To the staff of Mayo Clinic, science is indeed an act of love.

To my awesome ancestors—Cleo & Bill, Wanda & Don, Paul &

Jean (and great-grandma Mary), and by marriage (JR, TK, Billie & Jack)—without your determination, unconditional love, and hard work, I wouldn't be here. I'm looking up to you in heaven from Big Sky Country and sending you so much love and gratitude.

APPENDIX

REFLECT

DISCOVER YOUR SERVICE SUPERPOWER

Envision your journey toward creating a better future. Think about your Service Superpower Profile strengths and your greatest concerns. Examples include: plastic pollution, lack of clean energy options, species extinction.

TIP: Set a five-minute timer to keep focused!

SET & SHARE GOALS

IDENTIFY YOUR TOP CONCERNS

When the five minutes are up, pause and reflect. If you are working with a group, listen carefully as you share your profiles and concerns.

TIP: Try creating your vision for a positive future.

DIVE IN & LEARN

RESEARCH & REGROUP

After setting goals, research ways to take action that support your top three concerns. Set a deadline and a follow-up meeting to share ideas for action. Remember, this is fun!

TIP: If you have trouble finding ways to help, check out www.onegreenthing.org for resources and ideas!

START TAKING ACTION

JOURNAL TO TRACK ACTIONS & JOY

The action plan will consist of your top three issues. Each day, take one action and then write down not only what you did but how it made you feel, using the Joy Tracker. Align your One Green Things with your profile.

TIP: Create a journal to reflect on the impacts of the actions.

CELEBRATE & LOOK AHEAD

WHAT WORKED? WHAT DIDN'T? WHAT'S NEXT?

Don't forget to celebrate at the end of the first month of action. Share how you are feeling. What actions made the biggest difference, what were simply fun, and what may have created the foundation for real change? Then ask yourself, What's next? Make a plan for your next thirty days and how to keep changing the world.

REMEMBER: Thirty days isn't enough for anyone to save the planet. Yet research shows that taking action reduces anxiety and helps create cultural change. Track your action. Recruit others to join you.

VOTE FOR CLIMATE ACTION

Support local, state, and federal legislators who support strong clean energy and climate policies.

Carbon emissions in the US saved by 2030: **3 gigatons**

REDUCE AIRLINE TRAVEL

Buy carbon offsets when you do.

Carbon emissions saved: **700 – 2800 kg/year**

GREEN YOUR PORTFOLIO

Check your retirement fund or savings account. Consider investing in renewables and divest from fossil fuels.

Carbon emissions saved: **A LOT!**
(and $14 trillion in the divest movement)

EAT MORE PLANTS, LESS MEAT

Reducing your meat intake by half can cut your carbon footprint by 40 percent.

Carbon emissions saved: **300 – 1600 kg/year**

WALK, BIKE, RIDE

Greening your commute by taking advantage of public transit or buying a more efficient car can also significantly reduce your footprint.

Carbon emissions saved living "car-free": **1000 – 5300 kg/year**

Carbon emissions saved with a more efficient car: **1190 kg/year**

Sources: David Suzuki Foundation, "Top 10 Things You Can Do About Climate Change"; Seth Wynes and Kimberly A. Nicholas, "The Climate Mitigation Gap: Education and Government Recommendations

REDUCE FOOD WASTE

Most families throw out 3 kg of otherwise edible food a week. Eliminating food waste or composting can make a big impact.

Carbon emissions saved: **530 kg/year**

SWITCH TO GREEN POWER

Call your energy provider to see if you can make the switch.

Carbon emissions saved: **1000 – 2500 kg/year**

FORGET FAST FASHION

The fashion industry contributes 10 percent of global carbon emissions. According to the thredUP fashion calculator, the average person's fashion footprint is 734.8 kg/year

Carbon emissions saved: **440 kg/year**

WASH IN COLD WATER

Washing laundry in cold water can result in significant energy savings.

Carbon emissions saved: **250 kg/year**

TALK ABOUT THE CLIMATE CRISIS

The biggest impact you can make is with your family, friends, and community. Talk about the need for big climate solutions. Reach out with compassion. Discuss being a good ancestor. Create a greener, heathier, more equitable world for the next generation.

Carbon emissions saved: **440 kg/year**

Miss the Most Effective Individual Actions"; Paul Hawken, Drawdown: The Most Comprehensive Plan Ever Proposed to Reverse Global Warming; thredUP Fashion Calculator.

JOY TRACKER

Use the tracker to connect feelings of joy (and maybe less anxiety!) to the habits you are building with One Green Thing.

To begin, write down your One Green Thing for each day of the week below:

DAY 1

DAY 2

DAY 3

DAY 4

DAY 5

DAY 6

DAY 7

Now think about how each activity made you feel. For each day, put a mark on the continuum to record how you felt overall—joyful, joyless, or somewhere in between.

JOY LESS — JOY FUL

JOY LESS — JOY FUL

JOY LESS — JOY FUL

JOY LESS — JOY FUL

JOY LESS — JOY FUL

JOY LESS — JOY FUL

JOY LESS — JOY FUL

ORGANIZATIONS TO EXPLORE BY SERVICE SUPERPOWER PROFILE

Adventurer

- Bahamas Plastic Movement (bahamasplasticmovement.org)
- City Kids Wilderness Project (citykidsdc.org)
- Latinos Outdoors (latinooutdoors.org)
- National Geographic Society (nationalgeographic.org)
- Outdoor Afro (outdoorafro.org)
- Sierra Club (sierraclub.org)
- Student Conservation Association (thesca.org)

Beacon

- Climate Reality Project (climaterealityproject.org)
- Earthjustice (earthjustice.org)
- Fair Fight (fairfight.com)
- Greenpeace (greenpeace.org)
- Green Latinos (greenlatinos.org)
- The Louisiana Bucket Brigade (labucketbrigade.org)
- Sunrise Movement (sunrisemovement.org)

Influencer

- Charity: Water (charitywater.org)
- Environmental Media Association (green4ema.org)
- Global Green (globalgreen.org)
- Indigenous Environmental Network (ienearth.org)
- Natural Resources Defense Council (nrdc.org)
- Plastic Pollution Coalition (plasticpollution.org)
- WE-ACT Justice (weact.org)

Philanthropist

- Deep South Center for Environmental Justice (dscej.org)
- Defenders of Wildlife (defenders.org)
- Give Green (givegreen.org)
- League of Conservation Voters (lcv.org)
- Moms Clean Air Force (momscleanairforce.org)
- Native Americans in Philanthropy (nativephilanthropy.org)
- Powershift Network (powershift.org)

Sage

- The Coalition on the Environment & Jewish Life (coejl.org)
- Evangelical Environmental Network (creationcare.org)
- Interfaith Power & Light (interfaithpowerandlight.org)
- Green Faith (greenfaith.org)
- Green the Church (greenthechurch.org)
- National Religious Coalition on Creation Care (nrccc.org)
- National Religious Partnership for the Environment (nrpe.org)

Spark

- National Environmental Education Foundation (neefusa.org)
- National Parks Foundation (nationalparks.org)
- National Wildlife Federation (nwf.org)
- Public Lands Alliance (publiclandsalliance.org)
- US PIRG (pirg.org) and the state PIRGs
- Local land trusts
- State-based environmental health organizations

Wonk

- Climate Central (climatecentral.org)
- Environmental Defense Fund (edf.org)
- Friends of the Earth (foe.org)

- The Nature Conservancy (nature.org)
- Project Drawdown (drawdown.org)
- Science Moms (sciencemoms.com)
- Union of Concerned Scientists (ucsusa.org)

ORGANIZATIONS TO EXPLORE BY AREA OF CHANGE

Know You Can't Go It Alone

Disaster relief organizations

- American Red Cross (redcross.org)
- Doctors Without Borders (doctorswithoutborders.org)
- CARE (care.org)
- UNICEF (unicef.org)
- World Central Kitchen (wck.org)

Think Beyond Your Age

Groups facilitating intergenerational conversation

- Elders Climate Action (eldersclimateaction.org)
- Fridays for Future (fridaysforfuture.org)
- Hip Hop Caucus (hiphopcaucus.org)
- Mothers Out Front (mothersoutfront.org)
- OneGreenThing (onegreenthing.org)
- Third Act (thirdact.org)
- Climate Mental Health Network (climatementalhealth.net)

See Energy in a New Light

Organizations pushing for climate justice & clean energy solutions

- 350 (350.org)
- CERES (ceres.org)
- Climate Justice Alliance (climatejusticealliance.org)
- EcoAmerica (ecoamerica.org)

- Grid Alternatives (gridalternatives.org)
- Indigenous Climate Action (indigenousclimateaction.com)
- Rocky Mountain Institute (rmi.org)

Understand That You Are What You Eat

Organizations focused on sustainable food

- Center for Science in the Public Interest (cspinet.org)
- Eat Real (eatreal.org)
- National Black Farmers Association (nationalblackfarmersassociation.org)
- National Sustainable Agriculture Coalition (sustainableagriculture.net)
- Growing Hope Globally (growinghopeglobally.org)
- Planting Justice (plantingjustice.org)
- United Farm Workers (ufw.org)

Protect the Source

Organizations focused on clean water and ocean protection for all

- American Rivers (americanrivers.org)
- Clean Water Action (cleanwateraction.org)
- Ocean Conservancy (oceanconservancy.org)
- The Oxygen Project (oxygenproject.org)
- Riverkeeper (riverkeeper.org)
- Surfrider Foundation (surfrider.org)
- Regional and local watershed groups

Vote with Your Wallet

Organizations fighting for environmental health

- Asian Pacific Environmental Network (apen4ej.org)
- Breast Cancer Prevention Partners (bcpp.org)
- Center for Environmental Health (ceh.org)

- David Suzuki Foundation (davidsuzuki.org)
- Environmental Working Group (ewg.org)
- Safer Chemicals, Healthy Families (saferchemicals.org)

Love Your Mother (Earth)

Organizations protecting wildlife and connecting people to nature

- Children & Nature Network (childrenandnature.org)
- Conservation International (conservation.org)
- Honor the Earth (honorearth.org)
- National Parks & Conservation Organization (npca.org)
- RARE (rare.org)
- The Wilderness Society (wilderness.org)
- Regional and local environmental organizations

Learn more about all these charities at candid.org or charitynavigator.org.

For suggestions on documentaries, films, books, and influencers to follow, visit www.onegreenthing.org/bookresources.

THE ONEGREENTHING NEWSLETTER

Every two weeks, receive the latest news on climate action and personalized One Green Things based on each Service Superpower. Sign up at www.onegreenthing.org.

For more inspiration, follow us on Instagram @onegreenthing and Twitter @onegreenthing

Sign up for my personal newsletter at www.heatherwhite.com.

NOTES

Introduction

1. Kevin J. Coyle and Lise Van Susteren, "The Psychological Effects of Global Warming on the United States: And Why the U.S. Mental Health Care System Is Not Adequately Prepared," National Wildlife Federation, February 2012, http://www.nwf.org/~/media/PDFs/Global-Warming/Reports/Psych_Effects_Climate_Change_Full_3_23.ashx.
2. Brooke Jarvis, "Teenagers at the End of the World," *New York Times Magazine*, July 21, 2020, https://www.nytimes.com/interactive/2020/07/21/magazine/teenage-activist-climate-change.html.
3. Matthew Ballew et al., "Which Racial/Ethnic Groups Care Most About Climate Change," Yale Center for Climate Change Communication, April 16, 2020, https://climatecommunication.yale.edu/publications/race-and-climate-change/.
4. Susan Clayton and Christie Manning, "Climate Change's Toll on Mental Health," American Psychological Association, March 29, 2017, https://www.apa.org/news/press/releases/2017/03/climate-mental-health; Sarah Jacquette Ray, *A Field Guide to Climate Anxiety* (Oakland: University of California Press, 2020).
5. Jonathan Watts and Denis Campbell, "Half of Child Psychiatrists Surveyed Say Patients Have Environment Anxiety," *The Guardian*, November 20, 2020, https://www.theguardian.com/society/2020/nov/20/half-of-child-psychiatrists-surveyed-say-patients-have-environment-anxiety; Fiona Harvey, "Anxious About Climate, 4 in 10 Young People Are Wary of Having Kids," *Mother Jones*, September 15, 2021, https://www.motherjones.com/environment/2021/09/global-survey-climate-change-anxiety-young-people-children-kids/.
6. "Any Anxiety Disorder," National Institute of Mental Health, accessed October 12, 2021, https://www.nimh.nih.gov/health/statistics/any-anxiety-disorder.shtml; Markham Heid, "Depression and Suicide Rates Are Rising Sharply in Young Americans, New Report Says. This May Be One Reason Why," *TIME*, March 14, 2019, https://time.com/5550803/depression-suicide-rates-youth/.
7. Sarah Kaplan and Emily Guskin, "Most American Teens Are Frightened by Climate Change, Poll Finds, and About 1 in 4 Are Taking Action," *Washington Post*, September 9, 2019, https://www.washingtonpost.com/science/most-american-teens-are-frightened-by-climate-change-poll-finds-and-about-1-in-4-are-taking-action/2019/09/15/1936da1c-d639-11e9-9610-fb56c5522e1c_story.html; John Zogby Strategies, *A Survey of Gen Z & Millennials' Behavior & Values*, United States Conference of Mayors (Washington, DC: US Conference of Mayors, 2020), 5, https://www.usmayors.org/wp-content/uploads/2020/01/USCM_National-Youth-Poll-FINAL.pdf.
8. Kate Julian, "Parenting Kids with Anxiety," *The Atlantic*, May 2020, https://www.theatlantic.com/magazine/archive/2020/05/childhood-in-an-anxious-age/609079/.

NOTES

9. Matthew Cappucci, "Australian Fires Had a Bigger Impact on Climate than Covid-19 Lockdowns in 2020," *Washington Post*, July 27, 2021, https://www.washingtonpost.com/weather/2021/07/27/australian-bushfires-smoke-climate-covid/; Oliver Whang, "We've Run Out of Hurricane Names. What Happens Now?," *National Geographic*, September 21, 2020, https://www.nationalgeographic.com/science/article/weve-run-out-of-hurricane-names-what-happens-now; Trevor Nace, "The Arctic Circle Hit 100°F Saturday, Its Hottest Temperature Ever," *Forbes*, June 22, 2020, https://www.forbes.com/sites/trevornace/2020/06/22/the-arctic-circle-hit-101f-saturday-its-hottest-temperature-ever/?sh=65cb67354eb6; Blacki Migliozzi et al., "Record Wildfires on the West Coast Are Capping a Disastrous Decade," *New York Times*, September 24, 2020, https://www.nytimes.com/interactive/2020/09/24/climate/fires-worst-year-california-oregon-washington.html.
10. Jonathan Watts, "We Have 12 Years to Limit Climate Change Catastrophe, Warns UN," *The Guardian*, October 8, 2018, https://www.theguardian.com/environment/2018/oct/08/global-warming-must-not-exceed-15c-warns-landmark-un-report.
11. Vicky McKeever, "Nearly Half of Young People Worldwide Say Climate Change Anxiety Is Affecting Their Daily Life," CNBC, September 14, 2021, https://www.cnbc.com/2021/09/14/young-people-say-climate-anxiety-is-affecting-their-daily-life.html.
12. Asha C. Gilbert, "Climate Change, Racism and Social Justice Concerns Affecting Gen Z's Physical and Mental Health," *USA Today*, April 21, 2021, https://www.usatoday.com/story/life/2021/04/21/climate-change-racism-and-social-justice-major-concerns-gen-z/7289512002/.
13. Laura Hampson, "Global heating' and "Eco-anxiety" among New Terms Added to Oxford English Dictionary, The Independent, October 22, 2021 (https://www.independent.co.uk/climate-change/sustainable-living/global-heating-oxford-english-dictionary-b1943302.html).
14. Anne Lamott, "12 Truths I Learned from Life and Writing," TED2017, TED, April 2017, https://www.ted.com/talks/anne_lamott_12_truths_i_learned_from_life_and_writing/up-next?referrer=playlist-incredibly_soothing_ted_talks.

Chapter I

1. Kevin J. Coyle and Lise Van Susteren, "The Psychological Effects of Global Warming on the United States: And Why the U.S. Mental Health Care System Is Not Adequately Prepared," National Wildlife Federation, February 2012, http://www.nwf.org/~/media/PDFs/Global-Warming/Reports/Psych_Effects_Climate_Change_Full_3_23.ashx.
2. Brooke Jarvis, "Teenagers at the End of the World," *New York Times Magazine*, July 21, 2020, https://www.nytimes.com/interactive/2020/07/21/magazine/teenage-activist-climate-change.html.
3. Matthew Ballew et al., "Which Racial/Ethnic Groups Care Most About Climate Change," Yale Center for Climate Change Communication, April 16, 2020, https://climatecommunication.yale.edu/publications/race-and-climate-change/.
4. Susan Clayton and Christie Manning, "Climate Change's Toll on Mental Health," American Psychological Association, March 29, 2017, https://www.apa.org/news/press/releases/2017/03/climate-mental-health; Sarah Jacquette Ray, *A Field Guide to Climate Anxiety* (Oakland: University of California Press, 2020).
5. Jonathan Watts and Denis Campbell, "Half of Child Psychiatrists Surveyed Say Patients Have Environment Anxiety," *The Guardian*, November 20, 2020, https://www.theguardian.com/society/2020/nov/20/half-of-child-psychiatrists-surveyed-say-patients-have-environment-anxiety; Fiona Harvey, "Anxious About Climate, 4 in 10 Young People Are Wary of Having Kids," *Mother Jones*, September 15, 2021, https://www.motherjones.com/environment/2021/09/global-survey-climate-change-anxiety-young-people-children-kids/.

6. "Any Anxiety Disorder," National Institute of Mental Health, accessed October 12, 2021, https://www.nimh.nih.gov/health/statistics/any-anxiety-disorder.shtml; Markham Heid, "Depression and Suicide Rates Are Rising Sharply in Young Americans, New Report Says. This May Be One Reason Why," *TIME*, March 14, 2019, https://time.com/5550803/depression-suicide-rates-youth/.
7. Sarah Kaplan and Emily Guskin, "Most American Teens Are Frightened by Climate Change, Poll Finds, and About 1 in 4 Are Taking Action," *Washington Post*, September 9, 2019, https://www.washingtonpost.com/science/most-american-teens-are-frightened-by-climate-change-poll-finds-and-about-1-in-4-are-taking-action/2019/09/15/1936da1c-d639-11e9-9610-fb56c5522e1c_story.html; John Zogby Strategies, *A Survey of Gen Z & Millennials' Behavior & Values*, United States Conference of Mayors (Washington, DC: US Conference of Mayors, 2020), 5, https://www.usmayors.org/wp-content/uploads/2020/01/USCM_National-Youth-Poll-FINAL.pdf.
8. Kate Julian, "Parenting Kids with Anxiety," *The Atlantic*, May 2020, https://www.theatlantic.com/magazine/archive/2020/05/childhood-in-an-anxious-age/609079/.
9. Annie Lowrey, "All That Performative Environmentalism Adds Up," *The Atlantic*, August 31, 2020, https://www.theatlantic.com/ideas/archive/2020/08/your-tote-bag-can-make-difference/615817/.
10. Tess Riley, "Just 100 Companies Responsible for 71% of Global Emissions, Study Says," *The Guardian*, July 10, 2017, https://www.theguardian.com/sustainable-business/2017/jul/10/100-fossil-fuel-companies-investors-responsible-71-global-emissions-cdp-study-climate-change; "Our Planet Is Drowning in Plastic—It's Time for Change!," United Nations Environment Program, July 2020, https://www.unep.org/interactive/beat-plastic-pollution/.
11. James Clear, *Atomic Habits: An Easy & Proven Way to Build Good Habits & Break Bad Ones* (New York: Avery, 2018), 17.
12. Clear, *Atomic Habits*, vii–viii.
13. Charles Duhigg, *The Power of Habit: Why We Do What We Do in Life and Business* (New York: Random House, 2014), 19–25.
14. Clear, *Atomic Habits*, 39.
15. BOTWC Staff, "50 Years Ago, Shirley Chisholm Was Sworn In as the First Black Congresswoman," Because of Them We Can, January 3, 2019, https://www.becauseofthemwecan.com/blogs/botwc-firsts/50-years-ago-today-shirley-chisholm-was-sworn-in-as-the-first-african-american-congresswoman.
16. "Children Interrupt BBC News Interview—BBC News," BBC News, March 10, 2017, YouTube video, 00:43, https://www.youtube.com/watch?v=Mh4f9AYRCZY&t=3s.
17. Scott Stump, "'Yes, I Was Wearing Pants': Dad Talks About Being Interrupted by Kids on Live TV," *Today*, March 14, 2017, https://www.today.com/parents/dad-whose-kids-crashed-bbc-interview-speaks-out-first-time-t109218.
18. Allison Aubrey, "Happiness: It Really Is Contagious," National Public Radio, December 5, 2008, https://www.npr.org/templates/story/story.php?storyId=97831171.
19. John Tierney, "Good News Beats Bad on Social Networks," *New York Times*, March 18, 2013, https://www.nytimes.com/2013/03/19/science/good-news-spreads-faster-on-twitter-and-facebook.html; Ruoyun Lin and Sonja Utz, "The Emotional Responses of Browsing Facebook: Happiness, Envy, and the Role of Tie Strength," *Computers in Human Behavior* 52 (November 2015): 29–38, https://www.ncbi.nlm.nih.gov/pmc/articles/PMC4710707/.
20. Jessica Cerretani, "The Contagion of Happiness," *Harvard Medicine*, Summer 2011, https://hms.harvard.edu/magazine/science-emotion/contagion-happiness.
21. Phillippa Lally et al., "How Are Habits Formed: Modelling Habit Formation in the Real World," *European Journal of Social Psychology* 40, no. 6 (July 16, 2009): 998–1009, https://onlinelibrary.wiley.com/doi/abs/10.1002/ejsp.674.

NOTES

Chapter 2

1. "What Is ENVIRONMENTAL EQUITY?," The Law Dictionary, accessed October 11, 2021, https://thelawdictionary.org/environmental-equity.

Chapter 3

1. James L. Oschman, Gaetan Chevalier, and Richard Brown, "The Effects of Grounding (Earthing) on Inflammation, the Immune Response, Wound Healing, and Prevention and Treatment of Chronic Inflammatory and Autoimmune Diseases," *Journal of Inflammation Research* 8 (March 2015): 83–96, https://www.ncbi.nlm.nih.gov/pmc/articles/PMC4378297/.

Chapter 4

1. Alex M. Wood, Jeffrey J. Froh, and Adam W. A. Geraghty, "Gratitude and Well-Being: A Review and Theoretical Integration," *Clinical Psychology Review* 30, no. 7 (November 2010): 890–905, https://doi.org/10.1016/j.cpr.2010.03.005.

Chapter 6

1. *Merriam-Webster's Dictionary*, s.v. "philanthropist," accessed June 27, 2021, https://www.merriam-webster.com/dictionary/philanthropist.

Chapter 7

1. K. K. Ottesen, "An Evangelical Scientist on Reconciling Her Religion and the Realities of Climate Change," *Washington Post Magazine*, March 2, 2021, https://www.washingtonpost.com/lifestyle/magazine/an-evangelical-scientist-on-reconciling-her-religion-and-the-realities-of-climate-change/2021/02/26/f757d1c2-40b7-11eb-8bc0-ae155bee4aff_story.html.

Chapter 8

1. Derek Sivers, "How to Start a Movement," TED2010, TED, February 2010, https://www.ted.com/talks/derek_sivers_how_to_start_a_movement.
2. Gretchen Rubin, "Secret of Adulthood: Enthusiasm Is a Form of Social Courage," *Gretchen Rubin* (blog), April 3, 2014, https://gretchenrubin.com/2014/04/secret-of-adulthood-enthusiasm-is-a-form-of-social-courage/.

Chapter 10

1. Christopher Flavelle, "Climate Change Tied to Pregnancy Risks, Affecting Black Mothers Most," *New York Times*, June 18, 2020, https://www.nytimes.com/2020/06/18/climate/climate-change-pregnancy-study.html.
2. Sarah Kaplan, "Climate Change Is Also a Racial Justice Problem," *Washington Post*, June 29, 2020, https://www.washingtonpost.com/climate-solutions/2020/06/29/climate-change-racism/.
3. Renee Cho, "Why Climate Change Is an Environmental Justice Issue," *State of the Planet*, Columbia Climate School, September 22, 2020, https://news.climate.columbia.edu/2020/09/22/climate-change-environmental-justice/; Nicola Jones, "How Native Tribes Are Taking the Lead on Planning for Climate Change," Yale Environment 360, February 11, 2020, https://e360.yale.edu/features/how-native-tribes-are-taking-the-lead-on-planning-for-climate-change; Daisy Simmons, "What Is Climate Justice?," Yale Climate Connections, July 29, 2020, https://yaleclimateconnections.org/2020/07/what-is-climate-justice/.
4. Sigal Samuel, "What We Owe to Future Generations," Vox, July 2, 2021, https://www.vox.com/future-perfect/22552963/how-to-be-a-good-ancestor-longtermism-climate-change.
5. Roman Krznaric, *The Good Ancestor: A Radical Prescription for Long-Term Thinking* (New York: The Experiment, 2020).

Chapter 11

1. Heather White, "Moving Away from Climate Anxiety: How to Create a Family Eco-Action Plan," MBG Planet, October 25, 2020, https://www.mindbodygreen.com/articles/how-to-make-eco-action-plan-with-your-family.
2. Annie Lowrey, "All That Performative Environmentalism Adds Up," *The Atlantic*, August 31, 2020, https://www.theatlantic.com/ideas/archive/2020/08/your-tote-bag-can-make-difference/615817/.
3. Daniel Carpenter, "Yes, Signing Those Petitions Makes a Difference—Even if They Don't Change Trump's Mind," *Washington Post*, February 3, 2017, https://www.washingtonpost.com/news/monkey-cage/wp/2017/02/03/yes-signing-those-petitions-makes-a-difference-even-if-they-dont-change-trumps-mind/.
4. Chad Frischmann and Crystal Chissel, "The Powerful Role of Household Actions in Solving Climate Change," Project Drawdown, October 27, 2021, https://drawdown.org/news/insights/the-powerful-role-of-household-actions-in-solving-climate-change.

Chapter 12

1. Capi Lynn, "A Desperate Rescue: A Father's Heartbreaking Attempt to Save His Family from a Raging Fire," *Salem Statesman Journal*, September 11, 2020, https://www.statesmanjournal.com/in-depth/news/2020/09/10/oregon-wildfires-santiam-fire-evacuations-leave-family-members-dead/5759101002; Alejandra Borunda, "The Science Connecting Wildfires to Climate Change," *National Geographic*, September 17, 2020, https://www.nationalgeographic.com/science/article/climate-change-increases-risk-fires-western-us.
2. "Son, Grandmother Die in Fire, Mother in Burn Center," GoFundMe.com, September 30, 2020, https://ie.gofundme.com/f/help-chris-and-angie-after-devastating-loss?utm_campaign=p_cp_url&utm_medium=os&utm_source=customer.
3. Laura Millan Lombrana, Hayley Warren, and Brian K .Sullivan, "Heat, Floods, Fires: Jet Stream Is Key Link in Climate Disasters," Bloomberg Green, July 22, 2021, https://www.bloomberg.com/news/features/2021-07-22/how-climate-change-impacts-the-jet-stream-and-your-weather.
4. Edward O'Brien, "Fire Ecology Professor Says Ecosystem in 'Uncharted Territory,'" Montana Public Radio, June 23, 2021, https://www.mtpr.org/post/fire-ecology-professor-says-ecosystem-uncharted-territory; Bill Tripp, "Our Land Was Taken. But We Still Hold the Knowledge of How to Stop Mega-Fires," *The Guardian*, September 16, 2020, https://www.theguardian.com/commentisfree/2020/sep/16/california-wildfires-cultural-burns-indigenous-people.
5. Kasha Patel, "Extreme Flooding to Increase as Temperatures Rise, Study Finds," *Washington Post*, September 13, 2021, https://www.washingtonpost.com/weather/2021/09/13/extreme-flooding-increase-climate-change/.
6. "The Climate Denial Machine: How the Fossil Fuel Industry Blocks Climate Action," The Climate Reality Project, September 5, 2019, https://www.climaterealityproject.org/blog/climate-denial-machine-how-fossil-fuel-industry-blocks-climate-action.
7. Jonathan Watts, "We Have 12 Years to Limit Climate Change Catastrophe, Warns UN," *The Guardian*, October 8, 2018, https://www.theguardian.com/environment/2018/oct/08/global-warming-must-not-exceed-15c-warns-landmark-un-report; Nina Chestney and Andrea Januta, "U.N. Climate Change Report Sounds 'Code Red for Humanity,'" Reuters, August 9, 2021, https://www.reuters.com/business/environment/un-sounds-clarion-call-over-irreversible-climate-impacts-by-humans-2021-08-09/.
8. Brené Brown, *Braving the Wilderness: The Quest for True Belonging and the Courage to Stand Alone* (New York: Random House, 2017), 63.
9. Clara Strauss et al., "What Is Compassion and How Can We Measure It? A Review of Definitions and Measures," *Clinical Psychology Review* 47 (July 2016): 15–27, https://doi.org/10.1016/j.cpr.2016.05.004.

10. Inna Schneiderman et al., "Oxytocin During the Initial Stages of Romantic Attachment: Relations to Couples' Interactive Reciprocity," *Psychoneuroendocrinology* 37, no. 8 (August 2012): 1277–85, https://www.ncbi.nlm.nih.gov/pmc/articles/PMC3936960/; Markus MacGill, "What Is the Link Between Love and Oxytocin?," Medical News Today, September 4, 2017, https://www.medicalnewstoday.com/articles/275795.
11. Tom Seymour, "Everything You Need to Know About the Vagus Nerve: Further Research and Considerations," Medical News Today, June 28, 2017, https://www.medicalnewstoday.com/articles/318128#Further-research-and-considerations; Nancy Eisenberg et al., "Relations of School Children's Comforting Behavior to Empathy-Related Reactions and Shyness," *Review of Social Development* 5, no. 3 (April 2006): 330–51, https://www.researchgate.net/publication/229462469_Relations_of_School_Children's_Comforting_Behavior_to_Empathy-Related_Reactions_and_Shyness.
12. Dacher Keltner, "The Compassionate Species," *Greater Good Magazine*, July 31, 2012, https://greatergood.berkeley.edu/article/item/the_compassionate_species; David DiSalvo, "Forget the Survival of the Fittest: It Is Kindness That Counts," *Scientific American*, February 26, 2009, https://www.scientificamerican.com/article/kindness-emotions-psychology/.
13. Keltner, "Compassionate Species."
14. DiSalvo, "Forget the Survival of the Fittest."
15. Kristen Neff, "What Is Self-Compassion?," Self-Compassion, accessed July 1, 2021, https://self-compassion.org/the-three-elements-of-self-compassion-2/#3elements.
16. Kristen Neff, "The Physiology of Self-Compassion," Self-Compassion, accessed July 1, 2021, https://self-compassion.org/the-physiology-of-self-compassion/.
17. Kristen Neff, "Self-Appreciation: The Flipside of Self-Compassion," Self-Compassion, accessed July 1, 2012, https://self-compassion.org/self-appreciation-the-flip-side-of-self-compassion/.
18. Allison Abrams, "Is a Mental Health Crisis the Next Pandemic?," *Psychology Today*, March 17, 2021, https://www.psychologytoday.com/us/blog/nurturing-self-compassion/202103/is-mental-health-crisis-the-next-pandemic.
19. Vicky McKeever, "Nearly Half of Young People Worldwide Say Climate Change Anxiety Is Affecting Their Daily Life," CNBC, September 14, 2021, https://www.cnbc.com/2021/09/14/young-people-say-climate-anxiety-is-affecting-their-daily-life.html.
20. Paul Hawken, *Drawdown: The Most Comprehensive Plan Ever Proposed to Reverse Global Warming* (New York: Penguin Books, 2017), 220–23.
21. Wendi C. Thomas, "Louisville's Experiment: Can Teaching Empathy Boost Math Scores?," *Christian Science Monitor*, October 5, 2016, https://www.csmonitor.com/EqualEd/2016/1005/Louisville-s-experiment-Can-teaching-empathy-boost-math-scores.
22. "Climate Change and Health," World Health Organization, October 30, 2021, https://www.who.int/news-room/fact-sheets/detail/climate-change-and-health.
23. "Each Country's Share of CO2 Emissions," Union of Concerned Scientists, updated August 12, 2020, https://www.ucsusa.org/resources/each-countrys-share-co2-emissions.
24. William Brangham and Murrey Jacobson, "A Leaked UN Report Warns, 'Worst Is Yet to Come,' on Climate Change. Here's How You Can Help," PBS, June 23, 2021, https://www.pbs.org/newshour/show/a-leaked-un-report-warns-worst-is-yet-to-come-on-climate-change-heres-how-you-can-help.

Chapter 13

1. Lucy Harvey, "For 50 Years, Days of Our Lives Has Made History. Now, It's Part of the Smithsonian," *Smithsonian Magazine*, November 9, 2015, https://www.smithsonianmag.com/smithsonian-institution/sands-hourglass-days-of-our-lifes-50-anniversary-180957213/.

2. Mary Oliver, "The Summer Day," in *Devotions: The Selected Poems of Mary Oliver* (New York: Penguin Books, 2017), 316.
3. Roman Krznaric, *The Good Ancestor: A Radical Prescription for Long-Term Thinking* (New York: The Experiment, 2020).
4. Lisa Zaval, Ezra M. Markowitz, and Elke U. Weber, "How Will I Be Remembered? Conserving the Environment for the Sake of One's Legacy," *Psychological Science* 26, no. 2 (February 2015): 231–36, https://journals.sagepub.com/doi/10.1177/0956797614561266.
5. Michael Sanders and Sarah Smith, "Can Simple Prompts Increase Bequest Giving? Field Evidence from a Legal Call Centre," *Journal of Economic Behavior and Organization* 125 (May 2016): 179–91, https://www.sciencedirect.com/science/article /abs/pii/S0167268116000044.
6. Amnesty International, "Climate Change Ranks Highest as Vital Issue of Our Time –Generation Z Survey," December 10, 2019, https://www.amnesty.org/en/latest /press-release/2019/12/climate-change-ranks-highest-as-vital-issue-of-our-time/; "Generation Alpha and Environmental Consciousness," McCrindle https://mccrindle .com.au/article/blog/generation-alpha-and-environmental-consciousness/ (accessed January 28, 2024).
7. Jonathan Watts and Denis Campbell, "Half of Child Psychiatrists Surveyed Say Patients Have Environment Anxiety," *The Guardian*, November 20, 2020, https:// www.theguardian.com/society/2020/nov/20/half-of-child-psychiatrists-surveyed-say -patients-have-environment-anxiety; Fiona Harvey, "Anxious About Climate, 4 in 10 Young People Are Wary of Having Kids," *Mother Jones*, September 15, 2021, https:// www.motherjones.com/environment/2021/09/global-survey-climate-change-anxiety -young-people-children-kids/.
8. Nouran Salahieh, "More that a Third of US Population, from the Midwest to East Coast, Under Air Quality Alerts from Canadian Wildfire Smoke, CNN, June 28, 20203, https:// edition.cnn.com/2023/06/27/us/canada-wildfire-smoke-great-lakes/index.html.
9. "2023: A Historic Year of U.S. billion-dollar weather and climate disasters, National Oceanic and Atmospheric Agency, January 8, 2024, https://www.climate.gov /news-features/blogs/beyond-data/2023-historic-year-us-billion-dollar-weather-and -climate-disasters (accessed January 28, 2024); "Climate and Weather Related Disasters Surge Five-Fold Over 50 Years, but Early Warnings Saving Lives–WMO report," United Nations, September 1, 2021, https://news.un.org/en/story/2021/09/1098662 (accessed January 28, 2024).
10. "Any Anxiety Disorder," National Institute of Mental Health, accessed October 12, 2021, https://www.nimh.nih.gov/health/statistics/any-anxiety-disorder.shtml; Markham Heid, "Depression and Suicide Rates Are Rising Sharply in Young Americans, New Report Says. This May Be One Reason Why," *TIME*, March 14, 2019, https://time.com /5550803/depression-suicide-rates-youth/.
11. U.S. Department of Health and Human Services, Surgeon General, "Youth Mental Health Advisory," June 2021, https://www.hhs.gov/sites/default/files/surgeon-general -youth-mental-health-advisory.pdf?null.
12. Sarah Helizberger, "Gen Z Is the Loneliest Generation Survey Reveals, but Working Can Help," CNBC, May 2, 2018, https://www.cnbc.com/2018/05/02/cigna-study-loneliness-is -an-epidemic-gen-z-is-the-worst-off.html; Douglas Nemecek, "2018 Cigna U.S. Loneliness Index: Survey of 20,000 Americans Examining Behaviors Driving Loneliness in the United States," accessed October 12, 2021, PDF, 6, https://www.multivu.com/players /English/8294451-cigna-us-loneliness-survey/docs/IndexReport_1524069371598-173525450 .pdf; "Loneliness and the Workplace: 2020 U.S. Report," Cigna, accessed October 12, 2021, PDF, 1, https://www.cigna.com/static/www-cigna-com/docs/about-us/newsroom/studies -and-reports/combatting-loneliness/cigna-2020-loneliness-factsheet.pdf; Ceylan Yeginsu, "U.K. Appoints a Minister for Loneliness," *New York Times*, January 17, 2018, https:// www.nytimes.com/2018/01/17/world/europe/uk-britain-loneliness.html.

13. Youth Climate Survey 2022, Blue Shield of California, https://s3.amazonaws.com/cms.ipressroom.com/347/files/20223/Blue+Shield+of+California+NextGetn+Youth+Climate+Survey+2022+Report_FINAL.pdf (accessed January 28, 2024).
14. Kate Julian, "Parenting Kids with Anxiety," *The Atlantic*, May 2020, https://www.theatlantic.com/magazine/archive/2020/05/childhood-in-an-anxious-age/609079/.
15. Stephanie Kaufman, MSW, "When Reassurance is Hurting Your Child More than Helping," Anxiety & Depression Association of America, August 14, 2018, https://adaa.org/learn-from-us/from-the-experts/blog-posts/consumer/when-reassurance-hurting-your-child-more-helping.
16. John Cleese, *The Human Face*, BBC Television, 2001, https://quotepark.com/quotes/1931443-john-cleese-im-struck-by-how-laughter-connects-you-with-peopl/.

Chapter 14

1. Sarah Kaplan and Brady Dennis, "Amid Summer of Fire and Floods, a Moment of Truth for Climate Action," *Washington Post*, July 24, 2021, https://www.washingtonpost.com/climate-environment/2021/07/24/amid-summer-fire-floods-moment-truth-climate-action/.
2. "The Greenhouse Effect," UCAR Center for Science Education, accessed October 11, 2021, https://scied.ucar.edu/learning-zone/how-climate-works/greenhouse-effect.
3. Holly Shaftel, ed., "Climate Change: How Do We Know?," NASA: Global Climate Change, updated September 28, 2021, https://climate.nasa.gov/evidence/.
4. Mark Lynas, Bejamin Z. Houlton, and Simon Perry, "Greater than 99% Consensus on Human Caused Climate Change in the Peer-Reviewed Scientific Literature," *Environmental Research Letters* 16, no. 11 (October 2021), https://doi.org/10.1088/1748-9326/ac2966.
5. "Headline Statements from the Summary for Policyholders," Intergovernmental Panel on Climate Change, August 9, 2021, PDF, 1, https://www.ipcc.ch/report/ar6/wg1/downloads/report/IPCC_AR6_WGI_Headline_Statements.pdf; Brad Plumer and Henry Fountain, "A Hotter Future Is Certain, Climate Panel Warns. But How Hot Is Up to Us," *New York Times*, August 9, 2021, https://www.nytimes.com/2021/08/09/climate/climate-change-report-ipcc-un.html.
6. Jennifer Marlon et al., "Yale Climate Opinion Maps 2020," Yale Program on Climate Change Communication, September 2, 2020, https://climatecommunication.yale.edu/visualizations-data/ycom-us/; see also Jonathan Watts, "Case Closed: 99.9% of Scientists Agree Climate Emergency Caused by Humans," *The Guardian*, October 19, 2021, https://www.theguardian.com/environment/2021/oct/19/case-closed-999-of-scientists-agree-climate-emergency-caused-by-humans.
7. John Zogby Strategies, *A Survey of Gen Z & Millennials' Behavior & Values*.
8. "6 Arguments to Refute Your Climate-Denying Relatives This Holiday," Earth Day, updated December 20, 2021, https://www.earthday.org/6-arguments-to-refute-your-climate-denying-relatives/.
9. Bruce Lieberman, "1.5 to 2 Degrees Celsius of Additional Global Warming: Does It Make a Difference?," Yale Climate Connections, August 4, 2021, https://yaleclimateconnections.org/2021/08/1-5-or-2-degrees-celsius-of-additional-global-warming-does-it-make-a-difference/; "How Do We Know That Humans Are the Major Cause of Global Warming?," Union of Concerned Scientists, updated January 21, 2021, https://www.ucsusa.org/resources/are-humans-major-cause-global-warming.
10. Benjamin Franta, "Early Oil Industry Disinformation on Global Warming," *Environmental Politics* 30, no. 4 (2021): 663–68, https://www.tandfonline.com/doi/pdf/10.1080/09644016.2020.1863703.
11. Benjamin Franta, "Global Warming: From Scientific Warning to Corporate Casualty," TEDx Talks, July 20, 2021, YouTube video, 12:06, https://www.youtube.com/watch?v=Mp1JGqp7YMI.

12. Neela Banerjee, Lisa Song, and David Hasemyer, "Exxon: The Road Not Taken," Inside Climate News, September 16, 2015, https://insideclimatenews.org/project/exxon-the-road-not-taken/.
13. Linda Villarosa, "Pollution Is Killing Black Americans. This Community Fought Back," *New York Times Magazine*, July 28, 2020, https://www.nytimes.com/2020/07/28/magazine/pollution-philadelphia-black-americans.html.
14. Susanne Benz and Jennifer Burney, "US-Wide, Non-White Neighborhoods Are Hotter than White Ones," Advancing Earth and Space Science, July 13, 2021, https://news.agu.org/press-release/us-wide-non-white-neighborhoods-are-hotter-than-white-ones/; "Disparities in the Impact of Air Pollution," American Lung Association, April 20, 2020, https://www.lung.org/clean-air/outdoors/who-is-at-risk/disparities.
15. Nicola Jones, "How Native Tribes Are Taking the Lead on Planning for Climate Change," Yale Environment 360, February 11, 2020, https://e360.yale.edu/features/how-native-tribes-are-taking-the-lead-on-planning-for-climate-change.
16. "The Climate Crisis—A Race We Can Win," United Nations, accessed October 12, 2021, https://www.un.org/en/un75/climate-crisis-race-we-can-win.
17. David Roberts, "How to Drive Fossil Fuels Out of the US Economy, Quickly," Vox, August 6, 2020, https://www.vox.com/energy-and-environment/21349200/climate-change-fossil-fuels-rewiring-america-electrify.
18. Chelsea Eakin, "The US Can Reach 90 Percent Clean Electricity by 2035, Dependably and Without Increasing Consumer Bills," Berkeley Public Policy, June 9, 2020, https://gspp.berkeley.edu/faculty-and-impact/news/recent-news/the-us-can-reach-90-percent-clean-electricity-by-2035-dependably-and-without-increasing-consumer-bills.
19. Corinne Le Quéré et al., "Temporary Reduction in Daily Global CO2 Emissions During the COVID-19 Forced Confinement," *Nature Climate Change* 10 (2020): 647–53, https://www.nature.com/articles/s41558-020-0797-x; "After Steep Drop in 2020, Global Carbon Dioxide Emissions Have Rebounded Strongly," IEA, March 2, 2021, https://www.iea.org/news/after-steep-drop-in-early-2020-global-carbon-dioxide-emissions-have-rebounded-strongly.
20. Paul Hawken, *Drawdown: The Most Comprehensive Plan Ever Proposed to Reverse Global Warming* (New York: Penguin Books, 2017), 220–23.
21. Tess Riley, "Just 100 Companies Responsible for 71% of Global Emissions, Study Says," *The Guardian*, July 10, 2017, https://www.theguardian.com/sustainable-business/2017/jul/10/100-fossil-fuel-companies-investors-responsible-71-global-emissions-cdp-study-climate-change; "Revealed: The 20 Firms Behind a Third of All Carbon Emissions," *The Guardian*, October 9, 2019, https://www.theguardian.com/environment/2019/oct/09/revealed-20-firms-third-carbon-emissions.
22. "Renewable Energy," Center for Climate and Energy Solutions, accessed October 12, 2021, https://www.c2es.org/content/renewable-energy; Pippa Stevens, "Biden Administration Outlines How Solar Could Be Nearly Half of Electricity Supply by 2050," CNBC, September 8, 2021, https://www.cnbc.com/2021/09/08/white-house-solar-should-be-nearly-half-of-electricity-supply-by-2050.html.
23. "The Technocratic Approach to Climate Change, Explained," Climate Nexus, accessed October 12, 2021, https://climatenexus.org/climate-change-us/politics-and-policy/climate-change-technocratic-approach/#text-2.
24. "Padding Big Oil's Profits," Taxpayers for Common Sense, February 2, 2020, https://www.taxpayer.net/energy-natural-resources/padding-big-oils-profits/; "World Energy Outlook," International Energy Agency, accessed October 12, 2020, https://www.iea.org/topics/world-energy-outlook.
25. McKinsey & Co, "The Inflation Reduction Act: Here's What's In It," October 24, 2022, https://www.mckinsey.com/industries/public-sector/our-insights/the-inflation-reduction-act-heres-whats-in-it.

26. "How Cap and Trade Works," Environmental Defense Fund, accessed October 12, 2021, https://www.edf.org/climate/how-cap-and-trade-works; Bianca Nogrady, "China Launches World's Largest Carbon Market: But Is It Ambitious Enough?," *Nature*, July 20, 2021, https://www.nature.com/articles/d41586-021-01989-7.
27. Congress of the United States, "Effects of a Carbon Tax on the Economy and the Environment," Congressional Budget Office, May 2013, https://www.cbo.gov/sites/default/files/cbofiles/attachments/44223_Carbon_0.pdf.
28. Jamie Alexander, "No Matter Where We Work, Every Job Is a Climate Job Now," TEDx Talks, January 21, 2121, YouTube video, 8:58, https://www.youtube.com/watch?v=97qTunx8HZg.
29. Heather White, "How to Support a Clean Energy Future in Your Neighborhood, Whether You Own or Rent," MindBodyGreen, November 2020, https://www.mindbodygreen.com/articles/how-to-support-clean-energy-whether-you-rent-or-own-your-home.
30. Scott Wilson, "Amid Shut-Off Woes a Beacon of Energy," *Washington Post*, January 1, 2020, https://www.washingtonpost.com/climate-solutions/2020/01/01/amid-shut-off-woes-beacon-energy/.
31. Jason Coughlin et al., "A Guide to Community Solar: Utility, Private, and Non-Profit Project Development," National Renewable Energy Lab, November 2020, https://www.nrel.gov/docs/fy11osti/49930.pdf.
32. "Feeling Powerless? Switch to Green Power," Earth Day Network, April 11, 2020, https://www.earthday.org/feeling-powerless-switch-to-green-power/.
33. "Find Green-E Certified," Green-E Certified, accessed October 12, 2021, https://www.green-e.org/certified-resources.
34. "Aviation," Carbon Independent, updated September 13, 2021, https://www.carbonindependent.org/22.html.
35. Kiah Treece, "The 6 Best Carbon Offset Programs of 2022," Treehugger, updated March 19, 2021, https://www.treehugger.com/best-carbon-offset-programs-5076458.

Chapter 15

1. Saied Toossi, "National School Lunch Program," US Department of Agriculture, Economic Research Service, updated September 7, 2021, https://www.ers.usda.gov/topics/food-nutrition-assistance/child-nutrition-programs/national-school-lunch-program/; Kari Hamerschlag and Christopher D. Cook, "How Greener School Lunches Can Help Fight Climate Change," Green Schools National Network, April 6, 2017, https://greenschoolsnationalnetwork.org/greener-school-lunches-can-help-fight-climate-change/.
2. Centers for Disease Control and Prevention, "School Nutrition," February 15, 2021, https://www.cdc.gov/healthyschools/nutrition/schoolnutrition.htm.
3. Staff, "Going to Bed Hungry," *Washington Post*, January 27, 2021, https://www.washingtonpost.com/nation/interactive/2021/covid-hunger-crisis/.
4. Amy Bentley, "Ketchup as a Vegetable: Condiments and the Politics of School Lunch in Reagan's America," *Gastronomica* 21, no. 1 (2021): 17–26, https://online.ucpress.edu/gastronomica/article/21/1/17/116213Ketchup-as-a-VegetableCondiments-and-the-Politics.
5. Jennifer E. Gaddis, "The Big Business of School Meals," *Phi Delta Kappan*, September 21, 2020, https://kappanonline.org/big-business-school-meals-food-service-gaddis/.
6. Gaddis, "Big Business."
7. Jordan Shlain and Nora LaTorre, *Impact Report 2020* (Richmond, CA: Eat REAL, December 2020), https://hnanp3kj45p2kxeqv1p0sris-wpengine.netdna-ssl.com/wp-content/uploads/2020/12/Eat-REAL-Impact-Report-2020.pdf.

8. Gaddis, "Big Business."
9. "Our Impact," Chef Ann Foundation, accessed October 12, 2021, https://www.chefannfoundation.org/what-weve-done/our-impact.
10. "Results," Edible Schoolyard New Orleans, accessed October 12, 2021, https://esynola.org/stories-impact/results/; "Project Giving Gardens," Captain Planet Foundation, accessed October 12, 2021, https://captainplanetfoundation.org/project-giving-gardens/.
11. "Sources of Greenhouse Gas Emissions," United States Environmental Protection Agency, accessed October 12, 2021, https://www.epa.gov/ghgemissions/sources-greenhouse-gas-emissions.
12. "What Is the Farm Bill?," National Sustainable Agriculture Coalition, accessed October 12, 2021, https://sustainableagriculture.net/our-work/campaigns/fbcampaign/what-is-the-farm-bill/.
13. "Specialty Crop Block Grants," National Sustainable Agriculture Coalition, updated July 2019, https://sustainableagriculture.net/publications/grassrootsguide/local-food-systems-rural-development/specialty-crop-grants/.
14. "What Is the Farm Bill?," National Sustainable Agriculture Coalition; Judith Schwartz, "Soil as Carbon Storehouse: New Weapon in Carbon Fight?," Yale Environment 360, March 4, 2014, https://e360.yale.edu/features/soil_as_carbon_storehouse_new_weapon_in_climate_fight.
15. "Conservation Choices: Soil Health Practices," United States Department of Agriculture, April 2017, https://www.nrcs.usda.gov/Internet/FSE_DOCUMENTS/nrcseprd1318196.pdf.
16. Arianne Callender and Brendan DeMelle, "Obstruction of Justice," Environmental Working Group, July 20, 2004, https://www.ewg.org/research/obstruction-justice.
17. Emma Hurt, "The USDA Is Set to Give Black Farmers Some Debt Relief. They've Heard That One Before," *All Things Considered* (podcast), June 4, 2021, https://www.npr.org/2021/06/04/1003313657/the-usda-is-set-to-give-black-farmers-debt-relief-theyve-heard-that-one-before.
18. Hurt, "USDA Is Set to Give."
19. Miles McEvoy, "Organic 101: What the USDA Organic Label Means," US Department of Agriculture, March 13, 2019, https://www.usda.gov/media/blog/2012/03/22/organic-101-what-usda-organic-label-means; Kendra Klein and Anna Lappé, "You Have Pesticides in Your Body. But an Organic Diet Can Reduce Them by 70%," *The Guardian*, August 11, 2020, https://www.theguardian.com/environment/commentisfree/2020/aug/11/pesticide-danger-organic-food-roundup-study.
20. M. Shahbandeh, "Total Area of Land in United States Farms from 2000 to 2020 (in 1,000 Acres)," Statista, August 18, 2021, https://www.statista.com/statistics/196104/total-area-of-land-in-farms-in-the-us-since-2000/.
21. Schwartz, "Soil as Carbon Storehouse."
22. "Ask Your US Congress Members to Support New National Heat Regulation Bill," United Farm Workers, accessed October 12, 2021, https://act.seiu.org/a/heat2021; Andrew Moriarty, "Immigrant Farmworkers and America's Food Production: 5 Things to Know," FWD.us, March 18, 2021, https://www.fwd.us/news/immigrant-farmworkers-and-americas-food-production-5-things-to-know/; Lena Brook and Juanita Constible, "Treat Farmworkers as Essential, Not Sacrificial," NRDC, September 14, 2020, https://www.nrdc.org/experts/lena-brook/treat-farmworkers-essential-not-sacrificial.
23. "Genetically Modified Organisms," *National Geographic*, accessed October 12, 2021, https://www.nationalgeographic.org/encyclopedia/genetically-modified-organisms/.
24. "Genetically Modified Organisms," *National Geographic*.
25. Carey Gillam, "Glyphosate Fact Sheet: Cancer and Other Health Concerns," US Right to Know, September 27, 2021, https://usrtk.org/pesticides/glyphosate-health-concerns/.
26. Gillam, "Glyphosate Fact Sheet."

27. Alessio Fasano, "Zonulin and Its Regulation of the Intestinal Barrier Function: The Biological Door to Inflammation, Autoimmunity, and Cancer," *Physiological Reviews* 91, no. 1 (January 2011): 151–75, https://pubmed.ncbi.nlm.nih.gov/21248165/; Anthony Samsel and Stephanie Seneff, "Glyphoaste, Pathways to Modern Diseases II: Celiac Sprue and Gluten Intolerance, *Interdisciplinary Toxicology* 6, no. 4 (December 2013): 159–84, https://www.ncbi.nlm.nih.gov/pmc/articles/PMC3945755/.
28. Britt E. Erickson, "Bayer to End Glyphosate Sales to US Consumers," *Chemical & Engineering News*, July 30, 2021, https://cen.acs.org/environment/pesticides/Bayer-end-glyphosate-sales-US/99/web/2021/07.
29. Meg Wilcox, "Organic Diets Quickly Reduce the Amount of Glyphosate in People's Bodies," Environmental Health News, August 11, 2020, https://www.ehn.org/glyphosate-organic-food-2646939278.html.
30. Donna Berry, "Preparing for GMO Labeling in 2022," Food Business News, January 21, 2021, https://www.foodbusinessnews.net/articles/17559-preparing-for-gmo-labeling-in-2022.
31. "Plant-Rich Diets," Project Drawdown, accessed October 12, 2021, https://drawdown.org/solutions/plant-rich-diets.
32. "Plant-Rich Diets," Project Drawdown.
33. "10 Common Climate-Damaging Foods," NRDC, infographic, accessed October 12, 2021, https://www.nrdc.org/sites/default/files/10-common-climate-damaging-foods-infographic.pdf.
34. "EWG's Consumer Guide to Seafood," Environmental Working Group, September 18, 2014, https://www.ewg.org/consumer-guides/ewgs-consumer-guide-seafood#.W4Rl-C2ZPBJ.
35. "Plant-Rich Diets," Project Drawdown; Sarah Taber, "Farms Aren't Tossing Perfectly Good Produce. You Are," *Washington Post*, March 8, 2019, https://www.washingtonpost.com/news/posteverything/wp/2019/03/08/feature/farms-arent-tossing-perfectly-good-produce-you-are/; "Explore Solutions to Food Waste," ReFED, accessed October 18, 2021, https://insights-engine.refed.com/solution-database?dataView=total&indicator=us-dollars-profit.
36. Deborah Blum, *The Poison Squad: One Chemist's Single-Minded Crusade for Food Safety at the Turn of the Twentieth Century* (New York: Random House, 2019); Richard Fisher, "How to Decode a Food Label," BBC Future, June 23, 2021, https://www.bbc.com/future/article/20210623-how-to-decode-a-food-label.
37. Jensen Jose, "Cutting the GRAS," Center for Science in the Public Interest, July 28, 2021, https://www.cspinet.org/news/blog/cutting-gras.
38. Jose, "Cutting the GRAS."

Chapter 16

1. Water Science School, "The Water in You: Water and the Human Body," May 22, 2019, https://www.usgs.gov/special-topic/water-science-school/science/water-you-water-and-human-body; NOAA, "Rivers and Streams," *National Geographic*, accessed October 18, 2021, https://www.nationalgeographic.org/topics/resource-library-rivers-and-streams/; "What Is a Watershed?," National Ocean Service, updated February 26, 2021, https://oceanservice.noaa.gov/facts/watershed.html.
2. Bret Lemoine, "Milwaukee Sewage Overflow, Beaches Closed," Fox News, August 9, 2021, https://www.fox6now.com/news/milwaukee-sewage-overflow-beaches-closed.
3. Perry Beeman, "Des Moines River 'Essentially Unusable' for Drinking Water Due to Algae Toxins," *Iowa Capital Dispatch*, August 26, 2020, https://iowacapitaldispatch.com/2020/08/26/des-moines-river-essentially-unusable-for-drinking-water-due-to-algae-toxins/.
4. James F. McCarty, "Harmful Algal Blooms Continue to Plague Lake Erie, Threaten Drinking Water, Fish, Pets," *The Plain Dealer*, updated January 11, 2019, https://www.cleveland.com/metro/2017/08/harmful_algal_blooms_continue.html.

5. McCarty, "Harmful Algal Blooms."
6. Connie Lin, "What Is Red Tide? Deadly Florida Bloom Has Killed 800 Tons of Fish Along the Gulf Coast," *Fast Company*, July 20, 2021, https://www.fastcompany.com/90657102/what-is-red-tide-deadly-florida-bloom-has-killed-800-tons-of-fish-along-the-gulf-coast; Angela Fritz, "How Climate Change Is Making 'Red Tide' Algal Blooms Even Worse," *Washington Post*, August 15, 2018, https://www.washingtonpost.com/news/capital-weather-gang/wp/2018/08/14/how-climate-change-is-making-red-tide-algal-blooms-even-worse.
7. Jennie Lyons and Sierra Sarkis, "'Larger-than-Average' Gulf of Mexico 'Dead Zone' Measured," National Oceanic and Atmospheric Administration, August 3, 2021, https://www.noaa.gov/news-release/larger-than-average-gulf-of-mexico-dead-zone-measured; Elliott Negin, "Ask an Expert: Reviving the Gulf of Mexico's Dead Zone," The Equation: Union of Concerned Scientists, September 11, 2020, https://blog.ucsusa.org/elliott-negin/reviving-the-gulf-of-mexicos-dead-zone/.
8. Bob Henson and Jeff Masters, "Central Europe Staggers Toward Recovery from Catastrophic Flooding: More than 200 Killed," Yale Climate Connections, July 21, 2021, https://yaleclimateconnections.org/2021/07/central-europe-staggers-toward-recovery-from-catastrophic-flooding-more-than-200-killed/.
9. Brinley Hineman et al., "Waverly Flooding Victims: Family and Friends Reflect on the Loved Ones Lost," *Tennessean*, updated September 7, 2021, https://www.tennessean.com/in-depth/news/2021/08/25/waverly-tennessee-flooding-victims/8244501002/.
10. Ramy Inocencio, "12 Dead in Flooded Subway Car as China Experiences Record Rainfall," CBS News, July 22, 2021, streaming video, 1:54, https://www.cbsnews.com/video/12-dead-in-flooded-subway-car-as-china-experiences-record-rainfall/.
11. "Atmospheric Moisture Increase," Climate Signals, accessed October 18, 2021, https://www.climatesignals.org/climate-signals/atmospheric-moisture-increase.
12. Sarah Kaplan, "How Climate Change Helped Make Hurricane Ida One of Louisiana's Worst," *Washington Post*, August 30, 2021, https://www.washingtonpost.com/climate-environment/2021/08/29/how-climate-change-helped-make-hurricane-ida-one-louisianas-worst/.
13. Henry Fountain, "The Western Drought Is Bad: Here's What You Should Know About It," *New York Times*, October 21, 2021, https://www.nytimes.com/article/drought-california-western-united-states.html; Abraham Lustgarten "40 Million People Rely on the Colorado River. It's Drying Up Fast," *New York Times*, August 27, 2021, https://www.nytimes.com/2021/08/27/sunday-review/colorado-river-drying-up.html.
14. "Hexavalent Chromium," National Toxicology Program, February 2018, PDF, 1, https://www.niehs.nih.gov/health/materials/hexavalent_chromium_508.pdf.
15. Florence Williams, "How a Bunch of Scrappy Marines Could Help Vanquish Breast Cancer," *Mother Jones*, May/June 2012, https://www.motherjones.com/environment/2012/05/camp-lejeune-marines-breast-cancer-florence-williams/.
16. Ami Schmitz and Kristina Krohn, "Men Say Their Breast Cancer Was Caused by Contaminated Water at Camp Lejeune," NBC News, February 22, 2013, https://www.nbcnews.com/nightly-news/men-say-their-breast-cancer-was-caused-contaminated-water-camp-flna1c8505683; Greg Barnes, "Marine Veterans Petition for Medical Health Registry for Camp Lejeune Toxic Water Victims," NC Health News, March 16, 2021, https://www.northcarolinahealthnews.org/2021/03/16/marine-veterans-petition-for-medical-health-registry-for-camp-lejeune-toxic-water-victims/.
17. David Zucchino, "The Few, the Proud, the Stricken," *Los Angeles Times*, August 26, 2009, https://www.latimes.com/archives/la-xpm-2009-aug-26-na-military-cancer26-story.html.
18. Angela Canterbury and Abby Evans, "Toxic Secrecy: The Marines Corps' Cover-Up of Water Contamination at Camp Lejeune," Project On Government Oversight, June 29, 2011, https://www.pogo.org/analysis/2011/06/toxic-secrecy-marine-corps-cover-up-of-water-contamination-at-camp-lejeune; Staff, "Study: Possible Link between

Camp Lejeune Male Breast Cancer and Pollutants," *Tampa Bay Times*, September 21, 2015, https://www.tampabay.com/news/military/study-possible-link-between-camp-lejeune-male-breast-cancer-and-pollutants/2246464/.

19. Nathaniel Rich, "The Lawyer Who Became DuPont's Worst Nightmare," *New York Times Magazine*, January 6, 2016, https://www.nytimes.com/2016/01/10/magazine/the-lawyer-who-became-duponts-worst-nightmare.html.
20. Nathaniel Rich, "The Lawyer Who Became DuPont's Worst Nightmare"; Cheryl Hogue, "DuPont, EPA Settle," *Chemical and Engineering News* 83, no. 51 (December 19, 2005), https://cen.acs.org/articles/83/i51/DuPont-EPA-Settle.html.
21. John Flesher, "3M Reaches $10.3 Billion Settlement Over Contamination of Water Systems and Forever Chemicals, Associated Press, June 22, 2023, https://apnews.com/article/pfas-forever-chemicals-3m-drinking-water-81775af23d6aeae63533796b1a1d2cdb.
22. "PFAS Explained," US Environmental Protection Agency, accessed December 17, 2021, https://www.epa.gov/pfas/pfas-explained; Christophe Haubursin and Mac Schneider, "How 'Forever Chemicals' Polluted America's Water," Vox, August 4, 2020, https://www.vox.com/videos/2020/8/4/21354034/pfas-forever-chemicals-water-north-carolina; Annie Sneed, "Forever Chemicals Are Widespread in U.S. Drinking Water," *Scientific American*, January 22, 2021, https://www.scientificamerican.com/article/forever-chemicals-are-widespread-in-u-s-drinking-water/.
23. Kimberly Kindy, "States Take Matters into Their Own Hands to Ban 'Forever Chemicals,'" Washington Post, June 5, 2023, https://www.washingtonpost.com/politics/2023/06/05/forever-chemicals-state-bans-pfas/.
24. Brian Montag et al, "PFAS in Drinking Water: EPA Proposes Historic New Regulation," National Law Review, March 17, 2023, https://www.natlawreview.com/article/epa-releases-proposed-pfas-drinking-water-standards-and-expands-list-regulated.
25. Sherri Kolade, "Clean Water Activist Little Miss Flint Raises Almost $500K for Communities," *Michigan Chronicle*, April 20, 2021, https://michiganchronicle.com/2021/04/20/clean-water-activist-little-miss-flint-raises-almost-500k-for-communities/; Mari Copeny, "The Flint Water Crisis Began 5 Years Ago. This 11-Year-Old Activist Knows It's Still Not Over, *Elle*, April 24, 2019, https://www.elle.com/culture/career-politics/a27253797/little-miss-flint-water-crisis-five-years/.
26. Valerie Volcovici, "EPA Awards $100 Million to Upgrade Flint Water System," Reuters, March 17, 2017, https://www.reuters.com/article/us-michigan-water/epa-awards-100-million-to-upgrade-flint-water-system-idUSKBN16O288.
27. Jim Erickson, "Five Years Later: Flint Water Crisis Most Egregious Example of Environmental Injustice, U-M Researcher Says," University of Michigan News, April 23, 2019, https://news.umich.edu/five-years-later-flint-water-crisis-most-egregious-example-of-environmental-injustice-u-m-researcher-says/.
28. National Center for Environmental Health, Centers for Disease Control & Prevention, "Prevent Children's Exposure to Lead," accessed December 18, 2021, https://www.cdc.gov/nceh/features/leadpoisoning/index.html.
29. Derek Robertson, "Flint Has Clean Water Now. Why Won't People Drink It?," *Politico Magazine*, December 23, 2020, https://www.politico.com/news/magazine/2020/12/23/flint-water-crisis-2020-post-coronavirus-america-445459; Brakkton Booker, "Ex-Michigan Gov. Rick Snyder and 8 Others Criminally Charged in Flint Water Crisis, NPR, January 14, 2021, https://www.npr.org/2021/01/14/956924155/ex-michigan-gov-rick-snyder-and-8-others-criminally-charged-in-flint-water-crisi.
30. *Adapt Now: A Global Call for Leadership on Climate Resilience* (Rotterdam, The Netherlands: Global Commission on Adaption, 2019), 31, https://gca.org/wp-content/uploads/2019/09/GlobalCommission_Report_FINAL.pdf.
31. "Nature-Based Solutions," World Wildlife Foundation, accessed October 19, 2021, https://wwf.panda.org/discover/our_focus/climate_and_energy_practice/what_we_do/nature_based_solutions_for_climate/.

32. Charles Duhigg, "That Tap Water Is Legal but May Be Unhealthy," *New York Times*, December 16, 2009, https://www.nytimes.com/2009/12/17/us/17water.html.
33. "Drinking Water Regulations," United States Environmental Protection Agency, accessed October 19, 2021, https://www.epa.gov/dwreginfo/drinking-water-regulations; Mae Wu, "The Safe Drinking Water Act Must Be Updated Now," Natural Resources Defense Council, July 28, 2020, https://www.nrdc.org/experts/mae-wu/safe-drinking-water-act-must-be-updated-now.
34. Wu, "Safe Drinking Water Act."
35. Sarah Kaplan, "By 2050, There Will Be More Plastics than Fish in the World's Oceans, Study Says," *Washington Post*, January 20, 2016, https://www.washingtonpost.com/news/morning-mix/wp/2016/01/20/by-2050-there-will-be-more-plastic-than-fish-in-the-worlds-oceans-study-says/; Maanvi Singh, "It's Raining Plastic: Microscopic Fibers Fall from the Sky in Rocky Mountains," *The Guardian*, August 13, 2019, https://www.theguardian.com/us-news/2019/aug/12/raining-plastic-colorado-usgs-microplastics; Matthew Green, "'Punch in the Gut' as Scientists Find Micro Plastic in Arctic Ice," Reuters, August 14, 2019, https://www.reuters.com/article/us-environment-arctic-plastic/punch-in-the-gut-as-scientists-find-micro-plastic-in-arctic-ice-idUSKCN1V41V2; Melissa Locker, "It's Snowing Plastic in the Arctic Now," *Fast Company*, August 15, 2019, https://www.fastcompany.com/90390654/its-snowing-plastic-in-the-arctic-now.
36. Doyle Rice, "Oh, Yuck! You're Eating About a Credit Card's Worth of Plastic Every Week," *USA Today*, June 13, 2019, https://www.usatoday.com/story/news/nation/2019/06/12/plastic-youre-eating-credit-cards-worth-plastic-each-week/1437150001/.
37. John Vidal, "The Plastic Polluters Won 2019—and We're Running Out of Time to Stop Them," *The Guardian*, January 2, 2020, https://www.theguardian.com/environment/2020/jan/02/year-plastic-pollution-clean-beaches-seas; Carroll Muffet et al., *Executive Summary: Plastic & Climate: The Hidden Costs of a Plastics Planet* (Washington, DC: Center for International Environmental Law, 2019), PDF, 1, https://www.ciel.org/wp-content/uploads/2019/05/Plastic-and-Climate-Executive-Summary-2019.pdf.
38. "Pollution and Hazards from Manufacturing," Ecology Center, accessed June 15, 2020, https://ecologycenter.org/plastics/ptf/report3/.
39. "22 Facts About Plastic Pollution, and 10 Things We Can Do About It," EcoWatch, April 7, 2021, https://www.ecowatch.com/ocean-plastic-guide-2653277768.html?&warehouse10x=1; Tala Schlossberg and Nayeema Raza, "The Great Recycling Con," *New York Times*, December 9, 2019, streaming video, 5:18, https://www.nytimes.com/2019/12/09/opinion/recycling-myths.html; "Pollution and Hazards from Manufacturing," Ecology Center, accessed June 15, 2020, https://ecologycenter.org/plastics/ptf/report3/.
40. Break Free from Plastic Pollution Act of 2020, S. 3263, 116th Congress, https://www.congress.gov/bill/116th-congress/senate-bill/3263.
41. Erin Brockovich, *Superman's Not Coming: Our National Water Crisis and What We the People Can Do About It* (New York: Pantheon, 2020), 61–98.
42. "Guide to Safe Tap Water and Water Filters," Food & Water Watch, February 16, 2016, https://www.foodandwaterwatch.org/2016/02/16/guide-to-safe-tap-water-and-water-filters/; "Five Reasons to Skip Bottled Water," Environmental Working Group, September 22, 2013, https://www.ewg.org/consumer-guides/five-reasons-skip-bottled-water; Joey Grostern, "Environmental Impact of Bottled Water 'Up to 3,500 Times Greater than Tap Water," *The Guardian*, August 5, 2021, https://www.theguardian.com/environment/2021/aug/05/environmental-impact-of-bottled-water-up-to-3500-times-greater-than-tap-water; "Water Filter Ratings," Consumer Reports, accessed October 19, 2021, https://www.consumerreports.org/cro/water-filters.htm.

NOTES

Chapter 17

1. "Body Burden: The Pollution in Newborns," Environmental Working Group, July 15, 2005, https://www.ewg.org/research/body-burden-pollution-newborns.
2. Karissa Kovner, "Persistent Organic Pollutants: A Global Issue, a Global Response," United States Environmental Protection Agency, accessed October 21, 2021, https://www.epa.gov/international-cooperation/persistent-organic-pollutants-global-issue-global-response.
3. Robert O. Wright and Rosalind J. Wright, "The Institute for Exposomic Research," Icahn School of Medicine at Mount Sinai, accessed October 19, 2021, https://icahn.mssm.edu/research/exposomic.
4. Theo Colborn, Dianne Dumanoski, and John Peterson Myers, *Our Stolen Future: Are We Threatening Our Fertility, Intelligence, and Survival? A Scientific Detective Story* (New York: Plume, 1997).
5. Linda Birnbaum, "State of the Science of Endocrine Disruptors," *Environmental Health Perspectives* 121, no. 4 (April 2013): a107, https://www.ncbi.nlm.nih.gov/pmc/articles/PMC3620755/.
6. "Dirty Dozen Endocrine Disruptors: 12 Hormone-Altering Chemicals and How to Avoid Them," Environmental Working Group, October 28, 2013, http://www.ewg.org/research/dirty-dozen-list-endocrine-disruptors.
7. Manoj Kumar et al., "Environmental Endocrine-Disrupting Chemical Exposure: Role in Non Communicable Diseases," *Frontiers in Public Health* (September 24, 2020), https://www.frontiersin.org/articles/10.3389/fpubh.2020.553850/full; Pamela D. Noyes et al., "The Toxicology of Climate Change: Environmental Contaminants in a Warming World," *Environment International* 35, no. 6 (August 2009): 971–86, https://pubmed.ncbi.nlm.nih.gov/19375165/; Kumar et al., "Environmental Endocrine"; Allison J. Crimmins et al., *The Impacts of Climate Change on Human Health in the United States: A Scientific Assessment* (Washington, DC: U.S. Global Change Research Program, 2016), 22, https://health2016.globalchange.gov/low/ClimateHealth2016_FullReport_small.pdf.
8. "The Frank R. Lautenberg Chemical Safety for the 21st Century Act," United States Environmental Protection Agency, June 22, 2016, https://www.epa.gov/assessing-and-managing-chemicals-under-tsca/frank-r-lautenberg-chemical-safety-21st-century-act.
9. Eric Lipton, "How the Trump Administration Pulled Back on Regulating Toxic Chemicals," Yale Environment 360, February 12, 2019, https://e360.yale.edu/features/how-trump-administration-has-pulled-back-on-regulating-toxic-chemicals; Britt E. Erickson, "Podcast: TSCA Was Reformed 4 Years Ago. Is the US Chemical Law Living Up to Expectations?," June 17, 2020, in Chemical and Engineering News's *Stereo Chemistry*, podcast, 17:52, https://cen.acs.org/policy/chemical-regulation/Podcast-TSCA-reformed-4-years/98/i24; Tracey Woodruff et al., "5 Years After TSCA Reform: Strengthening Health Protection Through Science," Program on Reproductive Health and Environment, June 25, 2021, YouTube video, 1:21:47, https://prhe.ucsf.edu/5-years-after-tsca-reform.
10. "Canned Foods," Center for Environmental Health, December 31, 2021, https://ceh.org/products/canned-foods/; "Endocrine Disruptors," National Institute of Environmental Health Sciences, July 12, 2021, http://www.niehs.nih.gov/health/topics/agents/endocrine/.
11. George Citroner, "Food Industry's Switch to Non-BPA Linings Still Poses Health Risks," Health Line, July 25, 2019, https://www.healthline.com/health-news/common-chemicals-in-plastics-linked-to-childhood-obesity#Bisphenol-disrupts-the-bodys-metabolism.
12. Michael Hawthorne, "Chemical Companies, Big Tobacco, and the Toxic Products in Your Home," Tribune Watchdog: Playing with Fire: *Chicago Tribune*, May 6, 2012, http://media.apps.chicagotribune.com/flames/index.html; Liza Gross, "Flame

Retardants in Consumer Products Linked to Health and Cognitive Problems," *Washington Post*, April 15, 2013, https://www.washingtonpost.com/national/health-science/flame-retardants-in-consumer-products-are-linked-to-health-and-cognitive-problems/2013/04/15/f5c7b2aa-8b34-11e2-9838-d62f083ba93f_story.html.

13. University of Toronto, "New Flame Retardants, Old Problems: Replacement Flame Retardants Present Serious Risks, Caution Scientists," Science Daily, October 22, 2019, https://www.sciencedaily.com/releases/2019/10/191022080726.htm.
14. Valerie J. Brown, "Metals in Lip Products—A Cause for Concern?," *Environmental Health Perspectives* 121, no. 6 (June 1, 2013): A196, http://dx.doi.org/10.1289/ehp.121-a196.
15. Elizabeth Gamillo, "Scientists Find Toxic 'Forever Chemicals' in More than 100 Popular Makeup Products," *Smithsonian Magazine*, June 22, 2021, https://www.smithsonianmag.com/smart-news/hold-blush-cosmetics-may-contain-toxic-forever-chemicals-180978036/.
16. Sonya Lunder, "Asbestos Kills 12,000–15,000 people per year in the U.S.," Abestos Nation of EWG Action Fund, accessed September 20, 2021, http://www.asbestosnation.org/facts/asbestos-kills-12000-15000-people-per-year-in-the-u-s/.
17. Roni Caryn Rabin, "Women with Cancer Awarded Billions in Baby Powder Suit," *New York Times*, updated July 27, 2021, https://www.nytimes.com/2020/06/23/health/baby-powder-cancer.html.
18. Sarah Yang, "Teen Girls See Big Drop in Chemical Exposure with Switch in Cosmetics," Berkeley News, March 7, 2016, http://news.berkeley.edu/2016/03/07/cosmetics-chemicals/.
19. Julianna Deardorff and Louise Greenspan, *The New Puberty: How to Navigate Early Development in Today's Girls* (New York: Rodale, 2015); Heather White, "Are You There, God? It's Me, Rebecca: Early Puberty a New Normal," *Ms. Magazine*, February 4, 2015, http://msmagazine.com/blog/2015/02/04/are-you-there-god-its-me-rebecca-early-puberty-a-new-normal/.
20. See "Red List," Campaign for Safe Cosmetics, accessed October 20, 2021, https://www.safecosmetics.org/get-the-facts/chemicals-of-concern/red-list/; Sally Wadyka, "What You Need to Know About Sunscreen Ingredients," Consumer Reports, updated May 22, 2019, https://www.consumerreports.org/sunscreens/what-you-need-to-know-about-sunscreen-ingredients/.
21. Katherine Derla, "Sunscreen Ingredient Threatens Marine Life: Here's How Oxybenzone Kills Coral Reefs," *Tech Times*, October 22, 2015, http://www.techtimes.com/articles/98181/20151022/sunscreen-ingredient-threatens-marine-life-heres-how-oxybenzone-kills-coral-reefs.htm; Caroline Picard, "What the Proposed New FDA Sunscreen Rules Could Mean for You," *Good Housekeeping*, June 10, 2020, https://www.goodhousekeeping.com/health/a26470685/fda-sunscreen-regulations/.
22. "80 Years of the Federal Food, Drug, and Cosmetics Act," Food and Drug Administration, July 11, 2018, https://www.fda.gov/about-fda/virtual-exhibits-fda-history/80-years-federal-food-drug-and-cosmetic-act.
23. Oliver Milman, "US Cosmetics Are Full of Chemicals Banned by Europe—Why?" *The Guardian*, May 22, 2019, https://www.theguardian.com/us-news/2019/may/22/chemicals-in-cosmetics-us-restricted-eu.
24. "Red List," Campaign for Safe Cosmetics.
25. "Support the Safer Beauty Bill Package," Breast Cancer Prevention Partners, accessed October 20, 2021, https://www.bcpp.org/take-action/support-the-safer-beauty-bill-package/.
26. Kristen Rogers, "Potentially Harmful Chemicals Used in Many Cosmetics Products Banned by California Governor," CNN, (October 11, 2023), https://www.cnn.com/2023/10/11/health/california-bans-26-cosmetic-chemicals-wellness/index.html
27. *Fourth National Report on Human Exposure to Environmental Chemicals: Executive Summary* (Washington, DC: Centers for Disease Control and Prevention, 2009), PDF,

https://www.cdc.gov/exposurereport/pdf/FourthReport_ExecutiveSummary.pdf; "Welcome to the Human Toxome Project," Environmental Working Group, accessed June 15, 2020, https://www.ewg.org/sites/humantoxome/.

28. Daniel Penny, "Is Your Beloved Outdoors Gear Bad for the Planet?," *GQ*, January 22, 2021, https://www.gq.com/story/outdoor-gear-pfas-study.
29. National Biomonitoring Program, "Phthalates Factsheet," Centers for Disease Control and Prevention, April 5, 2021, https://www.cdc.gov/biomonitoring/Phthalates_FactSheet.html; "Get the Facts: Phthalates," Safer Chemicals, Healthy Families, accessed December 31, 2021, https://saferchemicals.org/get-the-facts/toxic-chemicals/phthalates/; Joseph M. Braun, Sheela Sathyanarayana, and Russ Hauser, "Phthalate Exposure and Children's Health," *Current Opinions in Pediatrics* 25, no. 2 (April 2013): 247–54, https://www.ncbi.nlm.nih.gov/pmc/articles/PMC3747651/.
30. "Household Chemical Products and Their Health Risk," Cleveland Clinic, May 24, 2018, https://my.clevelandclinic.org/health/articles/11397-household-chemical-products-and-their-health-risk.
31. Cliff Weathers, "5 Toxic Household Products You Probably Use Every Day," Salon, June 22, 2015, https://www.salon.com/2015/06/22/5_toxic_household_products_you_probably_use_every_day/.
32. "How Much Do Our Wardrobes Cost to the Environment?," World Bank, September 23, 2019, https://www.worldbank.org/en/news/feature/2019/09/23/costo-moda-medio-ambiente.
33. "How Much Do Our Wardrobes Cost to the Environment?"
34. UN Climate Change, "UN Helps Fashion Industry Shift to Low Carbon," United Nations Framework Convention on Climate Change, September 6, 2018, https://unfccc.int/news/un-helps-fashion-industry-shift-to-low-carbon; Christine Ro, "Can Fashion Ever Be Sustainable?," BBC Future, March 10, 2020, https://www.bbc.com/future/article/20200310-sustainable-fashion-how-to-buy-clothes-good-for-the-climate.
35. Deborah Drew and Genevieve Yehounme, "The Apparel Industry's Environmental Impact in 6 Graphics," World Resources Institute, July 5, 2017, https://www.wri.org/insights/apparel-industrys-environmental-impact-6-graphics; UN Climate Change, "UN Helps Fashion Industry Shift to Low Carbon," United Nations Framework Convention on Climate Change, September 6, 2018, https://unfccc.int/news/un-helps-fashion-industry-shift-to-low-carbon; Ro, "Can Fashion Ever Be Sustainable?"
36. Bella Webb, "Fashion and Carbon Emissions: Crunch Time," *Vogue Business*, August 26, 2020, https://www.voguebusiness.com/sustainability/fashion-and-carbon-emissions-crunch-time.
37. "Circular Economy: Definition, Importance and Benefits," European Parliament, updated March 3, 2021, https://www.europarl.europa.eu/news/en/headlines/economy/20151201STO05603/circular-economy-definition-importance-and-benefits.
38. Victoria Gilchrist and Heather White, "How to Talk About Racial Justice in Sustainability," GreenBiz, December 16, 2020, https://www.greenbiz.com/article/how-talk-about-racial-justice-sustainability.

Chapter 18

1. "History of Bison Management in Yellowstone," National Park Service, February 12, 2021, https://www.nps.gov/articles/bison-history-yellowstone.htm.
2. Wildlife Species Information, "American Buffalo (*Bison bison*)," US Fish and Wildlife Service, accessed October 20, 2021, https://www.fws.gov/species/species_accounts/bio_buff.html.
3. Wildlife Species Information, "American Buffalo (*Bison bison*)."
4. Dacher Keltner, Richard Bowman, and Harriet Richards, "Exploring the Emotional State of 'Real Happiness.' A Study into the Effects of Watching Natural History Television

Content," University of California, Berkeley, accessed October 20, 2021, https://asset-manager.bbcchannels.com/workspace/uploads/bbcw-real-happiness-white-paper-final-v2-58ac1df7.pdf; Bryan E. Robinson, "Why Watching Wildlife Programs Can Reduce COVID-19 Anxiety," *Psychology Today*, April 10, 2020, https://www.psychologytoday.com/us/blog/the-right-mindset/202004/why-watching-wildlife-programs-can-reduce-covid-19-anxiety.

5. Sasha Petrova, "Cat Lovers Rejoice: Watching Online Videos Lowers Stress and Makes You Happy," The Conversation, June 18, 2015, https://theconversation.com/cat-lovers-rejoice-watching-online-videos-lowers-stress-and-makes-you-happy-43460; Jessica Gall Myrick, "Emotion Regulation, Procrastination, and Watching Cat Videos Online: Who Watches Internet Cats, Why, and to What Effect?," *Computers in Human Behavior* 52 (November 2015): 168–76, https://www.sciencedirect.com/science/article/abs/pii/S0747563215004343.
6. Robb Q. Telfer, "De-Extinction Counter," Habitat 2030, September 9, 2016, https://habitat2030.org/blog/extinction/.
7. Bill Chappell and Nathan Rott, "1 Million Animals and Plant Species Are at Risk of Extinction, U.N. Report Says," May 6, 2019, on *All Things Considered*, NPR, podcast, 3:40, https://www.npr.org/2019/05/06/720654249/1-million-animal-and-plant-species-face-extinction-risk-u-n-report-says.
8. Rachel Nuwer, "Mass Extinctions Are Accelerating, Scientists Report," *New York Times*, June 2, 2020, https://www.nytimes.com/2020/06/01/science/mass-extinctions-are-accelerating-scientists-report.html.
9. John H. Cushman Jr., "After 'Silent Spring,' Industry Put Spin on All It Brewed," *New York Times*, March 26, 2001, https://www.nytimes.com/2001/03/26/us/after-silent-spring-industry-put-spin-on-all-it-brewed.html.
10. Corry Westbrook, "Endangered Species Act by the Numbers," National Wildlife Federation, PDF, 1, https://www.nwf.org/~/media/pdfs/wildlife/esabythenumbers.ashx; "Endangered Species," National Wildlife Foundation, accessed October 18, 2021, https://www.nwf.org/Educational-Resources/Wildlife-Guide/Understanding-Conservation/Endangered-Species.
11. Özgün Emre Can, Neil D. Cruze, and David W. MacDonald, "Dealing in Deadly Pathogens: Taking Stock of the Legal Trade in Live Wildlife and Potential Risks to Human Health," *Global Ecology and Conservation* 17 (January 2019): e00515, https://www.sciencedirect.com/science/article/pii/S2351989418302312.
12. One Health, "Zoonotic Diseases," Centers for Disease Control and Prevention, updated July 1, 2021, https://www.cdc.gov/onehealth/basics/zoonotic-diseases.html; Vanda Felbab-Brown, "Preventing Pandemics Through Biodiversity Conservation and Smart Wildlife Trade Regulation," Brookings Institution, January 25, 2021, https://www.brookings.edu/research/preventing-pandemics-through-biodiversity-conservation-and-smart-wildlife-trade-regulation/; Julie Shaw, "Why Is Biodiversity Important?," Conservation International, updated May 17, 2021, https://www.conservation.org/blog/why-is-biodiversity-important.
13. Corry Westbrook, "Endangered Species Act by the Numbers," 3.
14. "Endangered Species Act by the Numbers," 3.
15. "What Is the Relationship Between Deforestation and Climate Change?," Rainforest Alliance, updated August 12, 2018, https://www.rainforest-alliance.org/insights/what-is-the-relationship-between-deforestation-and-climate-change/.
16. Xiya Liang et al., "Research Progress of Desertification and Its Prevention in Mongolia," *Sustainability* 13, no. 12 (June 17, 2021): 6861, https://www.mdpi.com/2071-1050/13/12/6861.
17. NOAA, "What Is Blue Carbon?," National Oceanic and Atmospheric Administration, updated November 24, 2021, https://oceanservice.noaa.gov/facts/bluecarbon.html.

18. Institute for Carbon Removal Law and Policy, "Fact Sheet: Nature-Based Solutions to Climate Change," American University, https://www.american.edu/sis/centers/carbon-removal/fact-sheet-nature-based-solutions-to-climate-change.cfm; Jessica Martel, "Best Carbon Offset Programs," Investopedia, updated May 25, 2021, https://www.investopedia.com/best-carbon-offset-programs-5114611.
19. Calum Neill, Janelle Gerard, and Katherine D. Arbuthnott, "Nature Contact and Mood Benefits: Contact Duration and Mood Type," *The Journal of Positive Psychology* 14, no. 16 (December 16, 2018): 756–67, https://www.tandfonline.com/doi/citedby/10.1080/17439760.2018.1557242?s; Florence Williams, *The Nature Fix: Why Nature Makes Us Happier, Healthier, and More Creative* (New York: W. W. Norton, 2017), 25–31; Magdalena M. H. E. van den Berg et al., "Autonomic Nervous System Responses to Viewing Green and Built Settings: Differentiating Between Sympathetic and Parasympathetic Activity," *International Journal of Environmental Research and Public Health* 12, no. 12 (December 14, 2015): 15860–74, https://www.ncbi.nlm.nih.gov/pmc/articles/PMC4690962/; Gwen Dewar, "12 Benefits of Outdoor Play (and Tips for Helping Kids Reap These Benefits)," Parenting Science, accessed December 31, 2021, https://parentingscience.com/benefits-of-outdoor-play/.
20. Amy Fleming, "'It's a Superpower': How Walking Makes Us Healthier, Happier and Brainier," *The Guardian*, July 28, 2019, https://www.theguardian.com/lifeandstyle/2019/jul/28/its-a-superpower-how-walking-makes-us-healthier-happier-and-brainier.
21. ee Works, "The Benefits of Environmental Education for K–12 Students," North American Association for Environmental Education, accessed October 18, 2021, https://naaee.org/eepro/research/eeworks/student-outcomes.
22. Heather White, *Connecting Today's Kids with Nature: A Policy Action Plan* (Reston, VA: National Wildlife Federation, 2008), https://www.nwf.org/~/media/PDFs/Campus-Ecology/Reports/CKN_full_optimized.ashx.
23. Congressional Research Service, "Federal Land Ownership: Overview and Data," February 21, 2020, https://sgp.fas.org/crs/misc/R42346.pdf.
24. Marguerite Holloway, "Your Children's Yellowstone Will Be Radically Different," *New York Times*, November 15, 2018, https://www.nytimes.com/interactive/2018/11/15/climate/yellowstone-global-warming.html.
25. "Carbon Streaming Announces First Carbon Credit Stream Investment into a Blue Carbon Project," Businesswire, May 17, 2021, https://www.businesswire.com/news/home/20210517005852/en/Carbon-Streaming-Announces-First-Carbon-Credit-Stream-Investment-Into-a-Blue-Carbon-Project.
26. Richard Louv, *Our Wild Calling: How Connecting with Animals Can Transform Our Lives—and Save Theirs* (Chapel Hill, NC: Algonquin, 2019).

Chapter 19

1. Helena Dore, "Climate Strike Draws Bozeman High Schoolers, Local Leaders," *Bozeman Daily Chronicle*, September 24, 2021, https://www.bozemandailychronicle.com/news/city/climate-strike-draws-bozeman-high-schoolers-local-leaders/article_ca98b1dd-2ef0-542e-aca7-ddda0acb2eb4.html.
2. Wim Thiery et al., "Intergenerational Inequities in Exposure to Climate Extremes," *Science* 374, no. 6564 (September 26, 2001): 158–60, https://www.science.org/doi/10.1126/science.abi7339.
3. Sarah Kaplan, "Today's Kids Will Live Through Three Times as Many Climate Disasters as Their Grandparents, Study Says," *Washington Post*, September 26, 2021, https://www.washingtonpost.com/climate-environment/2021/09/26/change-disasters-kids-science-study/.

ABOUT THE AUTHOR

Heather White is the founder of OneGreenThing. White's twenty-plus years of service include working as a litigator at a prominent Nashville law firm, a campaign staffer and recount attorney for Al Gore's 2000 presidential campaign, the energy and environmental policy legislative counsel to United States Senator Russ Feingold, and an adjunct law professor at Georgetown University Law Center. White directed environmental education advocacy at the nation's largest conservation organization, ran an environmental health watchdog in Washington, DC, and led the nonprofit partner to Yellowstone National Park. She's a frequent spokesperson on environmental issues and has been featured on *Good Morning America*, CBS, PBS, NBC, and quoted in the *New York Times*, *Teen Vogue*, and *The Guardian*.

A native of East Tennessee, she loves country music, apple butter, and hiking. White received her bachelor's degree in environmental science from the University of Virginia, where she was an Echols Scholar. She earned her juris doctor from the University of Tennessee College of Law. She lives in Bozeman, Montana, with her husband and two teenage daughters.

www.heatherwhite.com
Twitter: @heatherwhiteofc
Instagram: @heatherwhiteofficial
Facebook: @heatherwhiteofficial